INDIAN CITIES

INDIAN CITIES
Ecological Perspectives

Edited by

VINOD K. TEWARI
JAY A. WEINSTEIN
V.L.S. PRAKASA RAO

CONCEPT PUBLISHING COMPANY, NEW DELHI

First Published 1986

Published by

Naurang Rai
Concept Publishing Company
H-13, Bali Nagar
NEW DELHI-110015 (India)

Printed in India by

Printway India
at S. Narayan & Sons
7117/18, Pahari Dhiraj, Delhi-110006

ACKNOWLEDGEMENTS

We acknowledge the contributors of the papers, and the publishers of the works already published who generously granted us permission to incorporate these works in this volume.

In the process of editing this book we have received institutional support from the Indian Institute of Management, Bangalore, Georgia Institute of Technology, Atlanta, and the American Studies Research Centre, Hyderabad. We are grateful to them.

We would also like to thank those who helped us in the production process, in particular Shri K.R. Karantha who typed the manuscript.

V.K. Tewari
J.A. Weinstein
V.L.S. Prakasa Rao

CONTENTS

LIST OF CONTRIBUTORS

Brian J.L. Berry

Director, School of Public Administration, Carnegie-Mellon University, Pittsburgh, Pennsylvania

John Brush

Professor of Geography at Rutgers University, New Brunswick, New Jersey

Noel P. Gist

Passed away in 1983. He was Professor of Sociology at the University of Missouri, Columbia, Missouri.

V.K. Pillai

Research Student in the Department of Urban Planning, University of Iowa, Iowa City, U.S.A.

V.L.S. Prakasa Rao

Honorary Professor at Centre for Economic and Social Studies, Hyderabad

Phillip H. Rees

(Formerly) Department of Geography, University of Chicago, Chicago

Arthur E. Smailes

Professor, Department of Geography, Queen Mary College, London, England

Howard E. Spodek

Professor of History at Temple University, Philadelphia, Pennsylvania

Vinod K. Tewari

Professor in Centre for Human-Settlements and Environmental Studies at the Indian Institute of Management, Bangalore

Jay A. Weinstein

Professor, School of Social Sciences at the Georgia Institute of Technology, Atlanta, U.S.A.

LIST OF FIGURES

LIST OF TABLES

ONE

VINOD K. TEWARI, JAY A. WEINSTEIN AND V.L.S. PRAKASA RAO

INTRODUCTION
An Overview

THE INDIAN CITY SYSTEM

THE city system in India which contains just over one-fifth of the nation's population dominates the country's economic and demographic profiles. The city system, through its industrial sector, contributes more to the national and international economy than to regional economies, resulting in a core-periphery dichotomy. In contrast, at the level of the city economy the bulk of the industrial and service labour force consists of middle and low income classes with low skills and low socio-economic status. To cater to this level of demand the "informal sector" is growing rapidly in all major cities. Both the million-plus cities and the "steel cities" with their highly capital intensive industries are partners in the creation of a "metropolitan-satellite" spatial structure. In spite of the industrial location policy in favour of small and medium sized towns, the cities continue to capture the major share of industrial growth.

The "Big Four"—Bombay, Calcutta, Delhi and Madras—are the leading partners. They play a major role in the organisation of economic space as the outlets of international trade as dependent enclaves and assembly points of primary produce from the hinterlands, with other cities acting as satellites to feed the metropolises. Thus the metropolitan economy, through its "top heavy" structure and enclave character, sustains an unbalanced and underdeveloped spatial economic structure (Raza *et al.* 1979).

The metropolises play the role of major distributors and redistributors of economic goods, and they dominate the primary sector, as evidenced, for example, by the excess of imports over exports of primary products in rural areas (Chakraborty 1978).

The million-plus cities, the "Big Four" particularly, dominate the hierarchical urban system, and they constitute the metropolitan apex with Bombay and Calcutta as the industrial and commercial metropolises. Bombay, with the largest stock exchange and the headquarters of Reserve Bank of India, functions as the financial capital while Delhi as the national capital has a distinct political role to play. Even among the "Big Four" there are functional gaps with Bombay, ranking first in the metropolitan functional scale followed by Calcutta, Delhi and Madras in that order.

India's urban population as recorded by the 1981 census was nearly 157 million. The urban population grew at the rate of about 5 per cent per year during the post-independent and plan period (1951-81). The number of million-plus cities agglomerations increased from one (Calcutta) in 1901 with a population of 1.5 million to twelve (Calcutta, Bombay, Delhi, Madras, Bangalore, Hyderabad, Ahmedabad, Kanpur, Pune, Nagpur, Lucknow and Jaipur) with a population of 41.6 million in 1981. In 1981, the population of these twelve cities was 27 per cent of the total urban population in the country, and grew at the rate of 4 per cent per year during 1971-81 (Table 1.1).

The city pull in population is more centripetal than centrifugal with sharp inter-regional imbalances in city distribution. The Indian cities particularly the million-plus cities have greater imbalance in sex-ratio varying from 698 females per 1000 males in Calcutta agglomeration to 917 in Hyderabad agglomeration (1971) indicating a male dominant migration. The converse of this is, there are more females than males in the source regions. This pattern is now changing with the migrants moving with their families.

CITY STRUCTURE

In its totality, a city is part of a regional system, which in its turn, is a part of a national urban system. Thus, a city system has intra-city, city-regional and inter-city dimensions, and both city-

TABLE 1.1 : **Indian cities with population of one million and more, 1981**

Urban agglomeration	Population in millions 1971	1981	Growth (%) 1971—1981
Calcutta	7.03	9.17	30.4
Greater Bombay	5.97	8.20	37.4
Delhi	3.65	5.23	43.4
Madras	3.17	4.28	34.9
Bangalore	1.65	2.91	76.2
Hyderabad	1.80	2.57	42.9
Ahmedabad	1.74	2.51	44.4
Kanpur	1.27	1.69	32.2
Pune	1.14	1.68	48.5
Nagpur	0.93	1.30	39.6
Lucknow	0.81	1.01	23.7
Jaipur	0.64	1.00	57.8
Total	28.35	41.55	39.5
All India Total Population	548.16	683.81*	24.7

* Includes projected population for Assam and J.K.
Source : Census of India, 1971 and 1981 (Provisional totals).

regional and inter-city dimensions have contributed to the shaping of the intra-city structure. Nevertheless, the papers presented in this volume are mainly concerned with the internal structure of India's cities. Therefore, the discussion is limited to the intra-city aspects only.

City structure is complex and comprises sets of entities—objects, activities, infrastructure and land—which are simultaneously interdependent and interactive among themselves and with their environment. The sub-systems of the structure, namely, social, economic, demographic, political and administrative are inter-related in a complex and probabilistic manner. To understand such a structure in its totality it is necessary that various components of the structure be analysed within a common conceptual framework. The study of city structure requires an understanding of the processes and patterns underlying the structure and the nature of the mechanisms involved in its creation and maintenance.

The questions examined in this collection of papers relate to the spatial patterning of population and activities within cities in India. It is generally the case that households with similar socio-economic status and similar land uses tend to cluster at various locations within a city. Actual locations are determined by a complex combination of historical, social, economic and administrative factors. This socio-geographic stratification changes with variations in the contributory factors that take place over a period of time. Thus, intra-city structure is dynamic. Sociologists, geographers, economists, and others have examined the social and economic characteristics of urban neighbourhoods in the search for generalizations about this dynamic spatial patterning. These studies, commonly known as "ecologies", seek to understand the relationships among human groups and between them and their socio-economic environments, first by describing how and, then, trying to explain why areas differ in the characteristics and behaviour of their human populations (Rees 1971). The major analytical approaches followed in these studies are morphological, social area, and factorial ecological.

Three basic "classical" models of spatial patterning—morphology—of households (concentric zone, sectoral and multiple nuclei) were originally formulated by Burgess (1928), Hoyt (1939), and Harris and Ullman (1945). Shevky and his associates (Shevky and Williams 1949, Shevky and Bell 1955), Mann (1965), and others have emphasised that the three patterns are not mutually exclusive and that the overall structure is tripartite or three-dimensional. According to Berry (1971), the three models are independent additive contributors to the total socio-economic structuring of urban neighbourhoods.

In his widely cited work, Sjoberg (1960) generalised the model with reference to "pre-industrial city" as one of "rich centre and poor periphery" with a single centre. Classifying it with all colonial cities in developing countries, McGee (1971) describes the structure of the Indian city as dual, with its "Bazaar-peasant economy" and "Firm type capitalist economy". Other morphological studies on Indian cities have emphasized this dual structure, with a traditional core and industrializing and "tertiarizing" belts or strips along the transport axes. Smailes (1969) refers to another type of dualism—indigenous and anglicised, with reference to morphological elements of the city.

In reality, the Indian city is far more complex than the concept of "dual structure" suggests. Cities with a history that predates the colonial era, such as Hyderabad, Varanasi and Delhi have exceedingly complicated structures characterized by a discontinuous telescoping of political, cultural and economic structures belonging to different historic periods. There is structural non-conformity, which is best expressed as "pre-industrial society" living in a "post-industrial city" with a slow rate of transformation (Ginsburg 1969). The Indian city—with its traditional elements, reminiscent of pre-industrial city and colonial city elements has also been characterised as a "collection of period pieces" (Jones 1966). Even the recent growth of Indian cities has been interpreted as a function of both pre-industrial fertility and post-industrial mortality (Davis, quoted by Dweyer 1974). In fact, in the early decades of this century, the Indian city was an "eater of men" (Spate and Learmonth 1967). But this is changing with a decline in mortality.

Some of the cities still retain their overall pre-industrial structural characteristics—if socio-economic status of the city's households is the determining criterion. Commercial cities like Vijayawada (population 3,17,000 in 1971), located at the apex of the agriculturally prosperous Krishna Delta on the east coast plain (Prakasa Rao and Ramachandran 1971), and Pollachi (population about 70,000), located in the rapidly commercialising groundnut tract of Coimbatore Plateau in the south central plateau of South India (Parvathi 1977) conform to the pre-industrial city model of 'rich centre—poor periphery'. Even as the periphery of these cities becomes richer with new residential extensions, the city centre still retains its economic prosperity and high socio-economic status.

When discussing structural patterns within an Indian city, it has become a ritual to echo that Western models—Burgess concentric zone model, Hoyt's sector model and Harris and Ullman's multiple nuclei model—are misfits. In this ritual, there is probably partial but not total truth. For one thing, the time specificity of the models does not seem to have been given due consideration. There seems to be a tendency to test the validity of the models mainly through examination of physical components, such as distribution of land use, or even economic ones, e.g., the distribution of income and occupational groups—without an appropriate analytical framework for structural analysis. The underlying assumptions of these models relate to a mono-centric city with the centripetal pull of the city centre, competition for space (or the role of accessibility), and functional inter-dependence, such as socio-economic status and retail trade structure in a locality, or functional exclusiveness (industrial zonation and higher income neighbourhoods). These need to be related to the persistent or emerging structural patterns, the underlying processes and, no less, to the presence (or absence) of degree and types of plan intervention. Differences between studies exist in research procedures, data base, levels of analysis and in the very objectives of study, not to mention the static analytical framework. With these limitations, total rejection of the structural patterns underlying the models is as unjustified as total acceptance of any given model. The search should continue for both conformities and deviations. The intra-city zonal patterns of million-plus cities, market cities, industrial cities and temple towns with their respective temporal and regional settings cannot be comprehended through one, universal, model of city structure.

OVERVIEW OF THE ESSAYS

The first ecological commentary on an Indian city in which the classic models are questioned is that of Gist (1957). The work was based on the earlier research in Latin America by Caplow (1949) and Hayner (1949). Scholarly interest in the internal structure and land use patterns of Indian cities is, however, as ancient as the cities themselves.

Gist, in his paper on "The ecology of Bangalore, India : an East-West comparison", has compared the ecological patterning of business, industry, public institutions, residential segregation and decentralization, and the slums in Bangalore city with the generalized patterns of these variables in American cities. He argues that while "ecological segregation in one form or another appears to be almost universal . . . the particular form in which such segregation occurs is highly variable, and changes in segregative patterns are affected by broad ideological, political, economic and technological changes." The data used in the study are based mainly on personal observations of the spatial distributions of the variables. Data from the 1941 and 1951 census have also been used. The spatial distributions are interpreted in terms of relative locations with respect to the two major retail and wholesale districts in the central part of the city. Gist observes that caste, education and occupational characteristics are correlated especially at the polar ends of the caste hierarchy. He acknowledges that ecological patterns in the non-Western cities differ from those of Western cities and using Bangalore as his case to represent the non-Western world he describes the various aspects of differentiation. Looking at the ecological changes in Latin American cities, he predicts a change in the structure of Bangalore in a direction of ecological configuration similar to that of West European or U.S. cities.

Smailes in his paper on the Indian city presents a descriptive model of the morphology of the city and identifies the dualism in the structure as a result of historical development of two distinct parts of the cities—the indigenous city and the British annexes. According to Smailes, the indigenous cities, whether they have grown around or alongside a fortress, royal palace or holy place, or have emerged from a village are irregular and amorphous. Compared with the native city, the annexes (cantonment, civil lines and railway colonies) were all, though in varying degree, organised and regular, spacious and sanitary, with buildings far more clearly distinguished and grouped according to their function. He identifies four morphological periods—first, the period of pre-urban nucleus ; second, the period of Moghul conquest and early European penetration (1550 to 1850) ; third, the period marked by the development of railways and by the establishment of the British Imperial Government (1850 to 1947) ; and fourth, the period of accelerated industrialization and extension of the transport system (since Independence in 1947).

Smailes further outlined the major morphological elements in Indian Cities as (i) old walled city, (ii) 'Modern' extramural extension, (iii) cantonment, (iv) civil lines, (v) railway colony, (vi) post-1947 estates and colonies. On an examination of the developments which are taking place in the cities after Independence, he concludes that the pace of urbanization in India has outstripped that of industrialization. Smailes presents a descriptive model of the Indian City formulated on the basis of personal observation but he does not employ a quantitative data base or related analytical techniques.

While the study by Gist describes the ecology of an Indian city (Bangalore) with its focus on a comparison of the city structure with that of a Western city, Smailes' study constructs a generalized model of the structure against the historical background of city growth and urbanization. Both the studies caution scholars of urban ecology against treating the generalizations about the genesis of urban settlements derived from assumptions and theories developed in the context of Western cities as having universal validity. Though the studies do not use quantitative data or statistical techniques, the descriptions of urban structure are analytical and giving critical insights.

Following these two descriptive essays a significant change in orientation can be discerned. The remaining papers examine the ecological structure of a number of Indian cities through the "factorial ecology" approach. This analytical approach is an improvement over the social area analysis, and is now an extensively used and accepted methodology for urban ecological analysis (see Berry ed. 1971 for a discussion and applications of factorial ecological approach). Factorial ecology approach was developed as an objective multi-variate analytical technique for the analysis of urban ecological structure. It uses factor analytic techniques to identify the underlying structural dimensions from a set of socio-economic characteristics selected for explaining the variation in the structure. Here the constructs are based on a wider set of socio-economic variables than those first used in social area analysis by Shevky and associates.

The first attempt of analysing the structure of an Indian city using the factorial ecology approach was made by Berry and Rees (1969) on Calcutta.

The purpose of the study was to examine the extent to which this archetypal "premature metropolis" is modernizing in a socio-geographic sense. The "actual" (i.e., objective) social structure and the geographic distribution of land use and populations of Calcutta were compared to the classic models, with the implicit understanding that the degree of fit between the model and reality is an indicator of degree of modernity.

In the Berry and Rees Calcutta study, principal factor analysis of data on 37 variables relating to family structure, literacy, type of employment, housing characteristics, and land use for 80 census wards of the city was used to operationalise the actual sociogeographic organisation of the city. The variables employed were taken from Census of India reports on population and from municipal land use surveys. According to the authors, Calcutta's structure is a mixture of pre-industrial and industrial ecologies, and is characterised by ethnic variability in the form of two zones, a concentric pattern of familism, and an axial pattern of areas according to degree of literacy. They also found both substantial and increasing geographic specialization of areas in business and residential land uses. The first part of the paper contains a good discussion on "pre-industrial" and "industrial" city structure, as well as a typology of social area analysis.

Although the study is based mainly on quantitative data and an objective methodology is used for the analysis, the results are not very clear. Particularly, the composition of the constructs of ecological dimensions, as judged by the factor loadings, is quite confused. This weakness stems in large part from the data base, that does not match with the requirements of ecological theory. Thus, the good intentions of bringing in objectivity in the analysis are of little value if one is not judicious in generating an appropriate data base.

Berry and Spodek, in their paper on Comparative Ecologies of Large Indian Cities, accept the fact that the data taken from the Census of India are not very appropriate for the study, and therefore, call their study a preliminary one. In this study, a comparative analysis of the ecologies of a number of large Indian cities—Ahmedabad, Bombay, Kanpur, Madras, Poona and Sholapur—is presented. While for the first four cities data are taken from 1961 census, for Poona and Sholapur survey data are used. Starting with a discussion of traditional residential patterns in Indian settlements, the authors analyse the principal dimensions in the factor structures of the cities for pattern recognition and comparison. They conclude that socio-economic status dimension is the most dominant of those accounting for the residential segregation in the cities, and the prevailing spatial pattern remains that of high status neighbourhoods in the core and low-status neighbourhoods at the periphery—substantiating Sjoberg's generalization about the pre-industrial city. The authors observe that "the increasingly diverse bases of social and economic power, which city life in modern India is generating, are transforming the structure". Thus, the structure is in a transitory stage, and generalisation of the model on which the emerging form will converge needs further work.

Brush, in his paper on Bombay and Delhi, has analysed changes in geographical patterns of growth in the two cities, based on factor analysis of 1961 and 1971 Census data. He demonstrates that there are inner zones of old upper-class concentration and outer sectors of recent upper-class expansion in both the cities. In addition, he finds that the urban cores retain features which reflect traditional preference of the elite for central locations, while recent peripheral growth parallels the pattern of Western cities.

The studies by Berry and Rees, Berry and Spodek, and Brush are based mainly on the Census of India data (except for the two cities of Poona and Sholapur), where the only variables which may have significant correlation with socio-economic status are literacy and Scheduled Caste population. By the Census definition, literates include all those who can read or write as well as professionally qualified persons. Occupational data are gathered by industrial classification of workers, and thus, have little or no relationship with status. However, it was established in these studies partly from the analysis of data and partly from the authors' perception of the cities, that socio-economic status dimension explains maximum variation in the residential patterns of Indian cities (Table 1.2).

This weakness of scanty data base in the research design was rectified in the studies on Madras, Ahmedabad, and Bangalore. These studies, in addition to census data, also incorporated data collected through primary survey, to generate appropriate variables for social area analysis.

In *Madras : An Analysis of Urban Ecological Structure in India*, Weinstein raises several questions about the assumptions of data gathering techniques and modes of interpretation employed in ecological research on Indian cities. He argues, like the authors of the other papers, that we have little or no warrant for suggesting that any of the Indian cities thus far studied is "modern" or "western" or even "industrial" by the criteria of social area analysis or factorial ecology. In support of this contention he uses social factor analysis to provide detailed evidence for the case of Madras. While Weinstein too employs

TABLE 1.2 : **Socio-economic status dimension in the factorial ecologies of Indian cities**

Study	Number of variables used in the study	Socio-economic status/ Social rank dimension: Status factor	Socio-economic status/ Social rank dimension: Variance explained
Berry and Rees (1969)			
Calcutta (1961)*	37	IV	**
Berry and Spodek (1971)			
Ahmedabad (1961)	29	I	17.4
Bombay (1961)	14	I	29.8
Kanpur (1961)	9	I	35.0
Madras (1961)	15	I	17.6
Poona (1954)	19	I	27.4
Sholapur (1965)	33	I	22.3
Weinstein (1974)			
Madras (1961)	14	I	24.5
Madras (1971)	14	I	22.3
Brush (1975)			
Delhi (1971)	18	I	17.4
Bombay (1971)	18	I	20.5
Weinstein and Vijayan (1978)			
Ahmedabad (1961)	12	I	18.8
Ahmedabad (1971)	12	I	37.9
Prakasa Rao & Tewari (1979)			
Bangalore (1974)	20	I	28.0

* Year against the city name refers to the reference year of the data used in the study.

**Not mentioned in the study.

1961 and 1971 census data, he supplements these with information gathered from a large-scale household survey and from personal observation. With qualifications, he concludes that concentric zone, sectoral and multiple nuclei models describe the actual structure of Madras no better than chance alone. At the same time, analysis of survey data provides strong support for earlier contentions that social rank—mixed, however, with "discrepent" characteristics such as presence of houseless persons—does dominate ecological structure in this city.

Weinstein and Vijayan's paper on Ahmedabad deals in some detail with the mixture of modernity and tradition in the ecological structure of Ahmedabad. Using historical, observational, census and survey data, the authors submit the proposition—first put forth by Berry and Spodek—that the ecological structure of Indian cities is "converging on the model industrial metropolis", to a relatively exact test. They find no such tendency in this "pre-colonial" walled city, but rather a complex mixture of trends toward both greater degree of modernity and greater degree of tradition.

Prakasa Rao and Tewari's study on the ecological structure of Bangalore City presents an analysis designed to examine to what extent the pattern of the ecological structure of the city follows the classical models. The study was based on household survey and census data. Twenty variables, representing social, economic, family, and migrant status-tenure-density dimensions were selected to analyse the degree of spatial variations and associated ecological patterning.

Bangalore's ecological structure was found to be not simple "rich centre and poor periphery" of the classical model, but a complex one : middle class city core and both high status and low status periphery. The outward moving periphery left behind older peripheries and constituted a distinct intermediate zone in the ecological scale. The polarised socio-economic groups were multi-nuclear in distribution ; concentric zonation was not mappable in simple geometric forms, but the dominance of the city core with its bi-focal structure, and distance from the city core were found to be significant in socio-economic zonal patterns. Sectoral patterns with mixed structure, industries and commercial establishments dominating, emerged along the transport arteries. Thus, the overall city structure was tripartite.

The last paper of this volume presents a theoretical statement on subjective components of urban ecological structure. In this paper, Weinstein traces the comparative history of the sub-field of subjective or "symbolic" ecology. This, much overlooked dimension—first stressed by Walter Firey in his pioneering study of Boston—promises to provide some answers to the question "why do Indian cities not conform to Western models?" In an explanatory analysis of previously collected census and survey data he suggests that virtually everywhere deviations from the models are found. Local cultural views, ethnic and caste loyalties, traditional segregation principles, particularistic cannons of neighbourhood desirability, and norms of intensive land use appear to be among the most prominent subjective elements influencing ecological structure in Indian cities.

CONCLUSIONS

The essays in this volume constitute what one might call the first chapter in the study of Indian city structure. For, classic ecology which has run its course and, now perhaps exhausted, can take its proper role as a technical adjunct—but no more than that—to urban research.

As social scientists, planners and others face the challenge of comprehending, anticipating and controlling urban growth in India, the concepts, assertions, and findings of this volume will surely prove to be valuable resources. But more than this, they will provide a caution about pursuing fruitless generalizations and about the limitations of heavy reliance on quantitative analysis of census (and even survey) data.

With this in view, this collection is offered not to promote sole or even principal reliance on factorial ecology in the study of Indian cities. Rather, as the volume as a whole suggests quite strongly, it is trusted that the work will inspire the beginning of a new stage in Indian urban research, in which methods and models are employed because there are reasons to believe that they will be helpful to solve real, practical problems of understanding and action—and to meet the complexity and enormous variety of Indian cities, with the open, empirical orientation and the entire range of technical and theoretical options which such a task demands.

TWO

NOEL P. GIST

THE ECOLOGY OF BANGALORE, INDIA*
An East-West Comparison

ECOLOGICAL research in American cities has revealed rather striking uniformities of ecological patterning, sufficient at least for certain tentative theories concerning urban ecology. It would be a fallacy, however, to assume that these theories are necessarily applicable to societies that differ strikingly from the United States in history, stage of technological development, socio-economic organization, and cultural interests.

Observations made in a few Latin American cities, for example, reveal ecological patterns quite different from those characteristic of many American cities (Caplow 1949). The classical urban pattern in these cities may be summarized as follows : high-status and high-income residents live near a central plaza, which is the social and institutional heart of the community ; low-status and low-income residents locate near the periphery ; economic establishments tend to be dispersed throughout the city, rather than highly centralized ; and suburban growth from residential decentralization is limited.

Ecological segregation in one form or another appears to be almost universal, but the particular form in which such segregation occurs is highly variable, and changes in segregative patterns are affected by broad ideological, political, economic, and technological changes. Probably all communities approaching

*Reprinted from *Social Forces*, Vol. 35, 1957, by permission of the authors and editors.

urban status also manifest some form of centralization of institutional functions. Such centralization usually reflects developments in communication, transportation, and economic organization, especially the growth of large bureaucratic structures. But it should not be assumed that Western forms of centralization have necessarily occurred in technologically and economically underdeveloped countries.

BANGALORE, ITS HISTORY AND GROWTH

From 1809 to the late 1940's the city was divided into two separate administrative units : Bangalore City on the west side, which was independently administered, and the Civil and Military Station (Cantonment), under British jurisdiction. With the merger of the two divisions after India's independence, the larger municipality assumed integrated administrative functions as the Bangalore Municipal Corporation.

Until the second decade of the present century the growth of Bangalore Municipality was comparatively slow, and the number of inhabitants was less than the number in the Cantonment. By 1921, however, the two administrative divisions were about equal in size (118,000 each). Thereafter the Municipality outdistanced the Cantonment in population growth. In the 1941 census the Municipality had a population of a quarter of a million compared with about 160,000 for the Cantonment. There is no doubt that the spatial arrangement of institutions and people was greatly influenced by this dual administrative organization and the presence of British military and civil personnel.

Since the terrain upon which Bangalore is located is comparatively level, without any sharp topographic contrasts to limit seriously or restrict the character of land usage, the ecological structure must, therefore, be interpreted mainly it terms of *social* factors that have operated to give the city its distinctive patterning. These factors have never functioned in a competitively *laissez faire* environment ; in recent times, at least, planning procedures have been systematically applied with the result that the city's ecology has been considerably affected. Large parks in the central part of the city, mainly in the Cantonment, are patently products of municipal planning.

But the scope and degree of planning have, nevertheless,

been limited. Certainly there is no rigidly planned economy, nor have tight restrictions been imposed specifying the location of economic establishments. Presumably most establishments related directly to the economy have been relatively free to compete for favourable sites and to locate without much interference from the government.

In 1941, before the merger, the Municipality had nearly 19,000 persons per square mile as compared with a density of about 12,000 persons in the Cantonment. Between 1901 and 1941, the metropolitan area increased by 35 per cent (from 9.8 to 13.2 square miles), but the population increased by 258 per cent and the density by 164 per cent. This rapid growth of population without a corresponding increase in occupied area has greatly accentuated the piling up of people in the metropolis. Actual increase in density has been considerable for all sections, but the range in density for different areas has been even more impressive. In 1941, the most sparsely settled section (which included a fashionable residential centre and the palace grounds) was 6,485 persons per square mile, but in the most densely settled area the corresponding density figure was 135,311 or twenty times as great.

THE ECOLOGY OF BUSINESS

Bangalore has a major retail and wholesale district, near the largest municipal market (Central Market), in the old Municipality, but the area has limited resemblance to the central business district of an American city. Central Market itself is a commercial nucleus, a centre of buying and selling, mainly of articles for household or personal consumption. Within the vicinity of the Market are streets lined with shops specializing in such merchandise as silk, cotton, and leather goods ; silverware, hardware, and musical instruments ; drugs, jewellery, and household furnishings ; supplies, books, and hemp products. In the district are also numerous small boarding and lodging hotels, pawnbrokers' establishments, lawyers' offices, and offices of various merchants' associations. Shops carrying the same type of merchandise tend to cluster together, creating small districts or sectors more or less specialized in character. One may find, for example, clusters of shops dealing with silverware or brassware or other clusters of shops selling bicycles or drugs.

This district is by no means limited exclusively to retail merchandising. As in American cities wholesale establishments are located conveniently near the retailers. Some of the retail and wholesale merchants receive certain kinds of merchandise directly from small-scale manufacturing enterprises located in the same area. On side streets, for instance, are weavers and ribbon manufacturers who supply nearby retail shops with finished articles. Certain merchants, notably dealers in indigenous drugs and silverware, perform dual functions of manufacturer and distributor, fabricating and selling commodities in the same building.

A wholesale district is located about one-fourth mile to the west of Central Market, on a major arterial thoroughfare and not far from the south-western boundary of the city. Wholesaling activity here is confined mainly to agricultural products such as grain, vegetables, ginger, garlic, and fruits, which are piled on the sides of streets or under shelters, to be disposed of to retailers. Even in this area open-air retail venders operate sidewalk businesses alongside the wholesalers. Hence there is no clear-cut areal differentiation based on the nature of distributive functions.

Nearly a mile north of Central Market, beyond the perimeter of the shopping district, is a "big business" street (Kempe Gowda Road) on which are concentrated banking and insurance firms, film distributing companies, offices of out-of-town newspaper correspondents, a large transportation firm, and the Bangalore Chamber of Commerce. Within a distance of about two or three blocks on this street are seven large motion picture houses, the largest concentration of theatres in the city.

About a half mile south of Central Market, and likewise beyond the major shopping district, is another area devoted mainly to banking and the financing and management of cooperatives. In this district there is a variety of cooperative institutions, including several cooperative banks, a land mortgage bank, a grain merchant's cooperative society, a house-building society, and government departments dealing with various cooperative enterprises.

Some two miles northeast of Central Market, in the Cantonment, are two important business areas. One of these, strung along the south side of Mahatma Gandhi Road for a distance of three or four blocks, and including portions of two intersecting streets, specializes mainly in luxury goods and services. Here are offices of airlines, movie theatres, photographers' establishments, expensive jewellery stores, department stores, large artcraft shops, book stores, radio and automobile sales houses, apparel shops, and pharmacies.

The other area, a few blocks north of Mahatma Gandhi Road, is adjacent to Russell Market, which is similar to, but smaller than, Central Market. The main shopping avenue in this area is Commercial Street, three blocks long, where almost everything, ranging from luxury items to household necessities, may be purchased. Many of the shops in this district are operated by Muslims, for the district itself is next to the principal Muslim residential section.

Ecological theory based on American studies has emphasized the importance of large-scale organizations, particularly business and industrial chains, in the centralization of functions in the central business district. Large corporate enterprises are commonly "headquartered" in the American central business zone although operations may be widely scattered. Furthermore, the department store has been an important feature of the central business district of the large or medium-sized city in the United States.

Most business in Bangalore is conducted in small shops. The chain-store type of merchandising organization has never developed to any extent, although a few enterprises, such as a small department store, a book store, and several banks, are links in regional chains of fairly small dimensions. Perhaps the nearest approach to the department store carrying products for mass consumption is the municipal market, dealing mainly in fruits, vegetables, flowers, hardware, inexpensive clothing, baskets and the like. But in the market place the stalls are privately managed. The nature of business is, therefore, not such as to foster the growth of complex bureaucracies so characteristic of the American metropolis. Hence the centralized office building housing an army of bureaucratic functionaries, including clerical staffs to do paper work, is almost non-existent.

The distribution of banking and insurance in Bangalore differs considerably from the prevailing patterns in large American cities. In this country banking and insurance tend to be rather highly centralized. There is decentralization, to be sure, but outlying banking institutions are commonly branch banks whose parent establishments are in the central business area. This also is the prevailing pattern for insurance, although there seems to be a trend toward decentralization.

In Bangalore, there are four principal clusters of large banks, with a few banking houses located somewhat apart from these clusters. None of these major banks is in the heart of the major business district, although there are numerous small-scale money-lending establishments in the area. One cluster of major banks is about a half mile north of Central Market ; another about a mile south of the Market ; a third in the Gandhi Road shopping area ; and a fourth in the retail district near Russell Market. The Imperial Bank of India, for example, is located two or three blocks south of Gandhi Road, in a spacious setting on the edge of an attractive residential district. Another, a comparatively small bank, British-owned, is located on the second floor of a building opposite the municipal offices, about a mile east of Central Market.

The distribution of hotels likewise represents a departure from the highly centralized pattern characteristic of American cities. There are numerous small hotels in the major business district and near the central railroad station, but the luxury hotels are some distance from concentrations of business or transportation facilities. For these hotels spacious and attractive surroundings appear to be more important considerations than proximity to business activities. No doubt attractiveness of location is important because many hotel residents are permanent or semi-permanent guests rather than transients. The West End Hotel, Bangalore's well known hostelry, is located at least a mile and a half from Central Market, and almost as far from the railway station, in a spacious semi-rural setting. The British apparently liked it this way.

The foregoing observations concerning the ecological patterning of business enterprises point up two facts relating to intercultural comparisons : First, ecological centralization of business is much more marked in the average American city than in Bangalore. Second, commercial areas of comparable size in an American city tend to be strikingly similar in their institutional characteristics, whereas each of the areas in Bangalore, described here briefly, is more or less specialized and, therefore, somewhat distinctive. This distinctiveness, born of specialization, is in part the result of Bangalore's political and military history. But the distinctiveness of business areas also reflects the differential buying power and cultural interests of the population. For example, British personnel resided mainly in the Cantonment, and the luxury goods enterprises in this area, especially those on Gandhi Road, were largely maintained by them.

THE ECOLOGY OF INDUSTRY

Although Bangalore is in the emergent stage of large-scale industrialization, much of the industrial output is still from small, independent industries, many of the cottage type. The smaller industries are widely scattered, mainly over the western part of the city, some of them near the railroad yards. As noted earlier, within the major business district, or in its vicinity, considerable manufacturing is carried on as an adjunct to the wholesale and retail establishments in the area. Indeed, manufacturing and selling is sometimes combined under one roof. A number of machine shops, including those with lathes and grinding equipment, have sprung up in the vicinity of Binny Mills, a large British-owned textile factory employing some seven thousand workers and located a short distance west of the principal business district. In this area also are builders and repairers of bullock carts and wheels, transportation vehicles used by wholesale dealers who are located close by. About a mile southeast of Central Market is the bamboo bazaar, and near it a slaughter house and a number of welding, forging, and weaving establishments.

But the larger industries tend to be rather highly decentralized. The extent to which decentralization of manufacturing has occurred depends on the kind and size of the particular industry. Three major textile mills are located near the western periphery of the city ; a large tobacco factory is situated on the northeastern edge. Others similarly located with reference to the peripheral settlement area include a porcelain factory, a lamp works, and brick and tile works, all on the west side. The major exception to this ecological pattern is the government soap factory, located in the central area occupied by public buildings.

Recent industrial developments are even more decentralized in location. Plants manufacturing airplanes and telephones are located five or six miles east of the municipal boundaries, and a factory producing electrical equipment is even farther removed on the west side. Henceforth all large-scale manufacturing will be located in a peripheral zone reserved by the Improvement Trust Board for industrial development (Fig. 2.1).

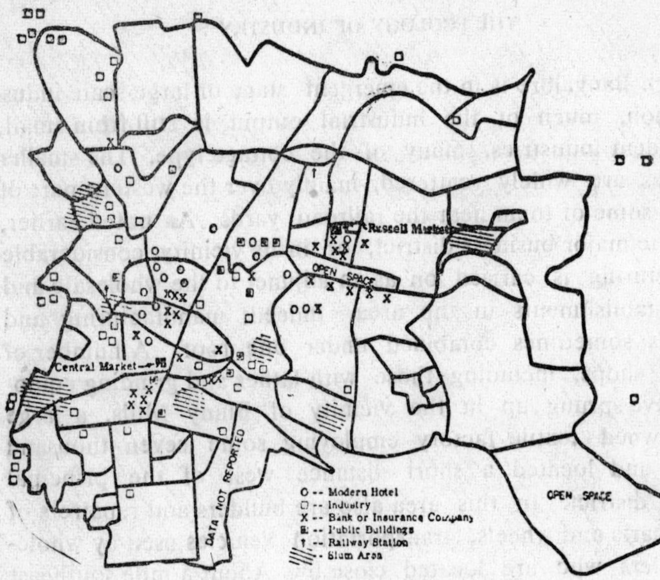

FIG. 2.1 : *Ecology of Bangalore's Institutions*

ECOLOGY OF PUBLIC INSTITUTIONS

In American cities most public buildings are usually located in or near the central business zone. Most of the public buildings in Bangalore are at some distance from the main centres of commercial activity. The municipal building (city hall), for example, is located on Narasimharaja Square, about a mile east of Central Market. Nearby is Town Hall (city auditorium). The main post office, the telephone and telegraph office, and the central police station are a mile and a half northeast of Central Market. Each of these organizations, of course, has suboffices in other sections of the city. The public library is in the centre of Cubbon Park, at least a mile from a business district.

Most of the city's major hospitals, on the other hand, are concentrated in a medical centre on the south side of Market Square, within hailing distance of Central Market. This appears to be the reverse of the pattern in American cities, where hospitals tend to be decentralized, or at least distributed rather widely over a metropolitan area. Clinics and dispensaries in Bangalore, however, are rather widely scattered, and there are a few hospitals removed from the major medical centre.

The location of public and semi-public institutions definitely indicates the role of municipal planning, especially in the Cantonment. A number of institutions are located within, or at the edge of, spacious Cubbon Park; these include the university, the city YMCA, Occupational Institute, the state government secretariat, and several cultural institutions. A mile or two south of Central Market, beyond the hospital area, is another cluster of cultural, religious, and educational institutions. Farther on, in Basavangudi, is the Indian Institute of Culture, a well known intellectual centre. Some institutions are removed from the city: Far to the northwest, beyond the municipal boundaries, is the Indian Institute of Science; to the southeast, also beyond the city, is a large mental hospital. Several agricultural experiment institutions are also located on the fringe.

Whether or not this planning represents the most efficient use of public institutions it is difficult to say. In some instances, such as the case of a public library set in the centre of a large park, it apparently does not.

RESIDENTIAL SEGREGATION

Urban ecologists have taken account of the tendency of people to select residential sites in cities on the basis of racial, cultural, religious, or ethnic preferences or prejudices, to choose residential locations that are symbolic of wealth, power, or social prestige, or to gravitate to low-income and low-prestige areas whose costs of occupancy are within the limits of their purchasing power. Ecological literature concerning American cities contains abundant examples of such segregative tendencies.

These tendencies are exhibited in Bangalore, as in western cities, except that caste preference or prejudice forms an additional basis for residential segregation. The importance of any of these factors depends, in the main, upon the values attached to them by particular individuals or families, although low purchasing power may override actual social preference in the selection of a residential location.

Viewing the ecological patterning of Bangalore in perspective, one may discern "natural areas" that are occupied predominantly or exclusively by members of a particular caste or religion and areas in which there is a mixed occupancy of people with different social affiliations and backgrounds. These segregative tendencies probably reflect the strength of communal sentiment, but they may also reflect prejudices against particular groups, especially low-caste or religious groups. By and large, the segregative tendency is probably stronger among the uneducated than among families on the higher income and educational levels where caste or religious interests and loyalties, if they exist, become secondary or even unimportant factors in selection of a residential site (Fig. 2.2).

Data are not available to indicate the residential locations of all Hindu castes, of which there are upwards of a hundred in the city. Even where a number of castes are considered collectively—"depressed" or "scheduled castes, for example—the broader categories conceal segregative tendencies of particular castes. Residential locations of "depressed" castes in the city, therefore, do not tell us anything much about the locations of, say, the Adikarnataka, Korama, and Koracha castes. But since the various depressed or untouchable castes have a social position and style of life that are fairly comparable, we are able to secure some idea of their ecological position in the community. In the accompanying map it is clear that the depressed castes manifest a tendency towards segregation, with some areas having more than half the residents in this class of castes. No doubt within these larger areas would be smaller areas in which occupancy by particular depressed castes would approach one hundred per cent (Fig. 2.3).

It is interesting to note from the map that areas of highest occupancy by depressed castes are on or near the city's outskirts. This is the reverse of the situation in the United States, where

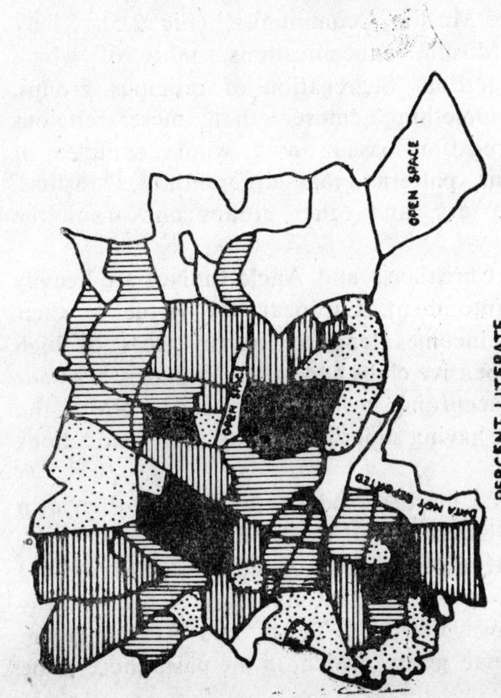

FIG. 2.2 : *Ecological pattern of per cent literates*

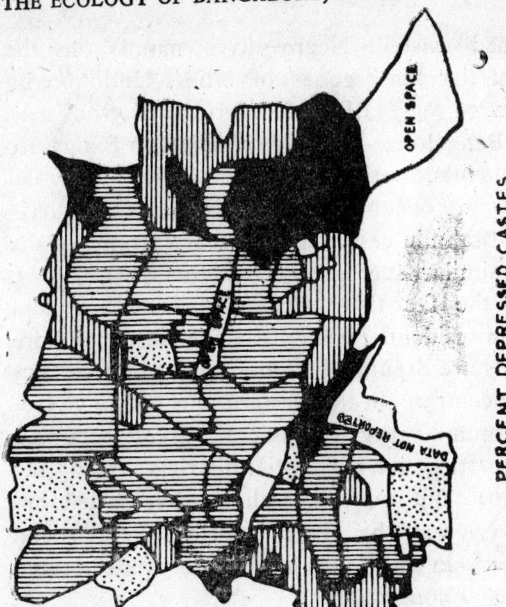

FIG. 2.3 : *Ecological pattern of per cent depressed castes*

the major "depressed caste"—Negro—lives mainly in the deteriorated sections of the inner zones of cities. Undoubtedly there is more pronounced segregation of American Negroes than of depressed castes in Bangalore. Only five districts in Bangalore had no low-caste residents, but most of the white residential areas in American cities are occupied exclusively by Caucasians.

Members of the Brahmin caste, at the other extreme of the social hierarchy, show similar segregative tendencies (Fig. 2.4). This is evidenced by the fact that 14 census districts have less than 1 per cent Brahmin residents, whereas in four districts more than half the inhabitants are Brahmins. There is a strong tendency for the Brahmins to concentrate in the western part of the city in the original Bangalore municipality. In general, areas occupied heavily by Brahmins are attractive residential districts, among the most attractive in the city. But not all Brahmins live in fashionable districts, whatever may be their position in the social hierarchy. Some of them, in fact, live in physically deteriorated houses under conditions of congestion.

Of the religious minorities in Bangalore none is perhaps as tightly segregated as the Muslim "community" (Fig. 2.5). There are four important Muslim concentrations, most of which represent extreme congestion. Segregation of religious groups, however, involves something more than mere religious differences ; it is segregation based on a whole complex of beliefs and behaviour patterns that differentiate Muslims, Hindus, Christians, Parsees, and other groups one from the other.

Europeans, Indian Christians, and Anglo-Indians are heavily concentrated in the Cantonment. Europeans by virtue of their comparatively superior incomes tend to live in areas of high quality houses, or at expensive clubs and hotels. Indian Christians are widely distributed according to their purchasing power, the affluent living in areas having superior housing accommodations (Fig. 2.6).

Anglo-Indians show a marked tendency to develop their own community life and, therefore, to reside in areas occupied by others of the same racial mixture. Most of them are concentrated in the southeastern portion of the Cantonment in districts locally known as Richmond Town, Langford Town, and Austin Town. As a culturally marginal group which, in the past, chose to be

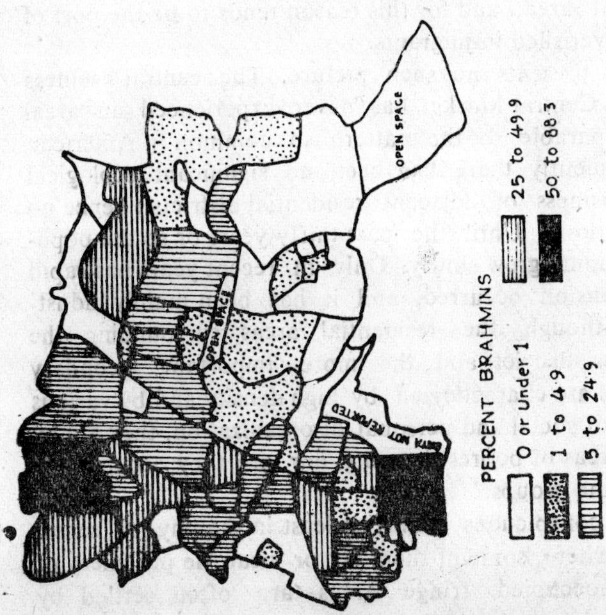

FIG. 2.4 : *Ecological pattern of per cent Brahmins*

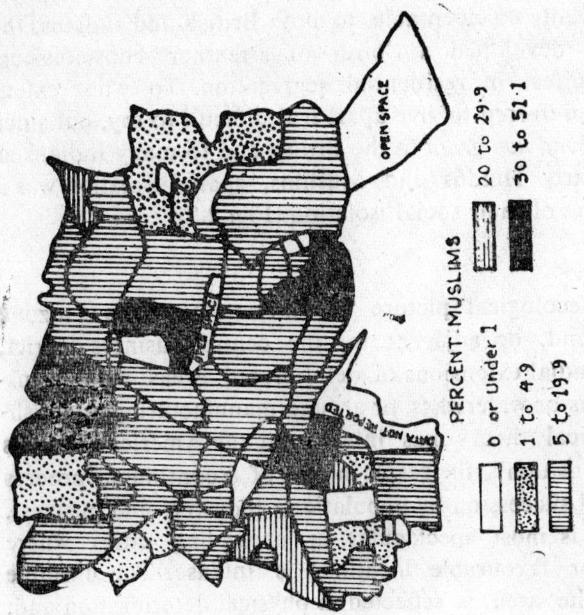

FIG. 2.5 : *Ecological pattern of per cent Muslims*

identified with Britain rather than with India, but whose members were often socially unacceptable to both British and Indians, the Anglo-Indians developed a "birds of a feather" consciousness which was manifest in residential segregation. To what extent they actually preferred to live apart it is difficult to say, but since they were *persona non grata* to the British and to many Indians as well, particularly Hindus and Muslims, spatial isolation was a logical outcome of their social isolation (Fig. 2.7).

The slum

The familiar ecological picture of an American slum is a zone centering around, or adjacent to, the central business district, with perhaps radial extensions of deteriorated areas along transportation routes or waterways, or around manufacturing establishments. Ecological theory has interpreted the American slum as an "area in transition", the result in part of a continuous process of invasion and succession by population, industry, and business. This invasion is most spectacular in expanding cities where competition for favourable locations is intense. The unstable character of the area is reflected in physical deterioration and, commonly, social disorganization. For the most part it is the city's low rent area, and for this reason tends to be the port of entry for impoverished in-migrants.

Bangalore presents no such picture. The central business district around Central Market has never experienced an areal expansion comparable to the pattern so common in American cities. Consequently there has been no significant ecological invasion by business of adjacent residential districts ; hence no "zone of transition". Until the past thirty years or so the population and economy grew slowly. Only in recent years has rapid economic expansion occurred, and it has been mainly industrialization. Although the residential areas surrounding the central business district and the more important secondary business centres are characterized by high density and by various manifestations of social and personal disorganization, they are by no means the areas of poorest housing, nor are they occupied by the lowest income groups.

The most conspicuous slums, at least in the physical sense, are located elsewhere, some of them on or near the periphery of Bangalore. Unoccupied fringe areas are often settled by

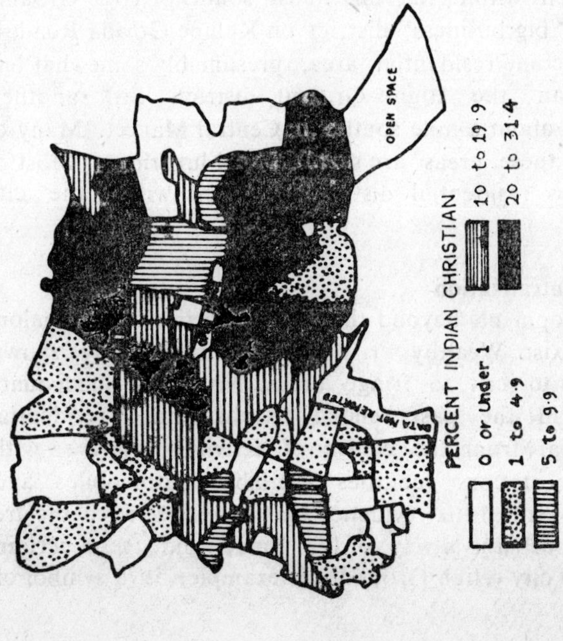

FIG. 2.6 : *Ecological pattern of per cent Indian Christian*

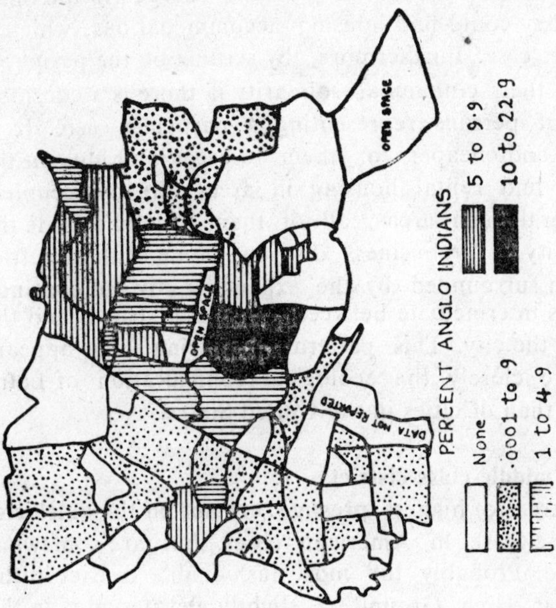

FIG. 2.7 : *Ecological pattern of per cent Anglo Indians*

in-migrants who cannot afford to pay rents charged in the inner zones, even if they could find housing accommodations, which is not always the case. Furthermore, by settling on the periphery they may retain their communal solidarity if there is a considerable migration of persons representing a particular caste. It is therefore easier and cheaper for them to erect mud huts on the fringe than to find rental housing in areas already occupied. There are several such areas, all of them located outside the immediate vicinity of the business districts. Some of these settlements have been surrounded by the expanding city but continue to exist at points intermediate between the outer fringe and the central zones of the city. This pattern of outlying slums appears to resemble more closely the ecological configuration of Latin American cities than of cities in the United States.

Upper-class and middle-class districts

The residential areas of highest prestige are not on the outskirts, as is usually the case in American cities, but are somewhat centrally located. Probably the most fashionable district in the city is on or near High Ground, a slightly elevated area in the northwest sector, about two miles north of the central business district but well within the city. Just south of High Ground, north of the "big business" district on Kempe Gowda Road, is another high-income residential area, presumably somewhat less fashionable than the High Ground district. Still another, Basavangudi, is about a mile south of Central Market. Many of the homes in these areas are spacious and luxurious. Most of the middle-class residential districts are well within the city limits.

Residential decentralization

Suburban developments beyond the settled portions of Bangalore simply do not exist. Wealthy or well-to-do families have shown little inclination to shift to fringe areas. No doubt inadequate transportation in the past tended to discourage residential location very far from the centre of activities, but even with modern transportation facilities available—automobiles and buses—there is still little evidence that Bangalore residents are interested in suburbia as a way of life. Since residence in certain areas within the city (High Ground, for example) is a symbol of prestige, high status families presumably prefer to be identified with residential localities that support or enhance their own claims to status. Certainly the suburban mode of living does not have the popular appeal it enjoys in the United States.

Nor have workers shown any marked tendency to develop industrial suburbs adjacent to the outlying manufacturing establishments. Even Hindustan Aircraft and the Indian Telephone Industries, located five or six miles beyond the city's boundaries, have not become the nuclei of satellite factory towns. The thousands of workers who are employed in these industries reside in Bangalore and commute by bus, work train, or bicycle. These industries are relatively new, however, and it may be that in time they will act as a magnet for suburban developments of working-men's homes. At any rate, the absence of any marked pattern of suburbanization gives the city a compactness that is not characteristic of most American metropolises.

This is not to say that fringe residential developments are completely absent. Public housing programmes, mainly for working-class or lower-white-collar families, are located on or near the city's edge. Housing construction under private auspices likewise tends to assume this pattern—extensions of the outer edges of metropolitan settlement rather than the development of detached satellite communities. Basavangudi, a new upper-class district on the south side of the city, is a case in point.

CONCLUSIONS

The present ecological structure of Bangalore bears the heavy imprint of the city's historical past. That past was characterized by limited industrialization, slow economic growth, divided political authority, and extensive planning, especially in the Cantonment. Today there is an upsurge of industrialism. Bangalore is the industrial boom town of South India. A planned economy is in prospect, both for the city and the nation as a whole. There will undoubtedly be continued population growth, and rapid growth at that.

What effect, if any, will these changes or others have on the basic ecological structure of the city ? Will industrialization, along with developments in trade, transportation, and communication, evolve a pattern similar to the western configuration ? Will communal segregation (caste and religion) decline in the wake of measures taken to strengthen political and economic democracy? Only a self-assured prophet would attempt to provide specific answers to these questions.

A possible clue to future ecological changes may be found in Latin American cities. Dotson (1954), Caplow (1949) and Hayner (1945) observed a tendency for the classical pattern to change under the impact of technology and North American or European value systems. The most industrialized metropolis of Latin America, Mexico City, has exhibited rather striking changes in the direction of ecological configurations similar to those of cities in the United States or western Europe. It is possible, then, that the impact of industrial teehnology, large-scale economic organization, modern transportation, and western status values will bring about similar changes in the ecological structure of Bangalore.

There is reason to believe that industry will become increasingly decentralized and that industrial districts will eventually emerge. In fact, the Improvement Trust Board has already delineated several fringe districts for new industrial developments, and all of these areas are well beyond the borders of the city. What is not so clear are possible trends in the ecology of business. Fundamental changes in the organization of trade, such as the growth of large-scale chain enterprises, would likely make for greater centralization of controls and hence stimulate the growth of a central business area. But such changes will likely come slowly. For a long time Bangalore will be a city of small shops.

Nor is it clear what the trend will be in patterns of residential segregation. The removal of caste barriers penalizing the untouchables or scheduled castes, together with a general improvement in their economic position, may make it possible for them, as was as other minority groups, to achieve a more favourable ecological location. But the ties of caste, family and religion are strong, and existing ecological patterns so deeply imbedded in custom, that changes will probably come slowly for most of the groups. Possibly the most effective force making for re-distribution of the various caste, ethnic, or religious groups will be slum clearance and public housing programmes in which occupants will be selected on the basis of need rather than social or cultural attributes.

The foregoing observations demonstrate clearly a fundamental problem of ecological theory as well as of sociological theory in general. It is the hazards involved in developing theories of society based upon observed data from a single society or culture area. Generalizations derived from a single culture tend to give American sociology a rather narrow provincialism which seriously limits its usefulness outside our national boundaries.

THREE

ARTHUR E. SMAILES

THE INDIAN CITY*
A Descriptive Model

THE peculiarly intimate association between cities and the civilisations to which they belong and of which they bear the stamp in their townscapes is obvious to the most casual observer. That variant forms of city and town should be among the most striking manifestations of culture differences is in the nature of things, for they are parts of the earth's surface where the products of human activity provide an especially heavy overlay and where transformation of the original landscape is peculiarly complete. Often, however, the expression of a culture in the townscape is appreciated only in its more superficial aspects, whereas in fact it involves the very structure of the city in terms of its basic components and their disposition. Assumptions and theories about city structure tend to be culture bound, and especially has it happened that generalisations derived from a study of western cities have been treated as of universal validity. As a corrective to such facile and false notions, salutary attention has been drawn by Sjoberg (1960) to fundamental differences between pre-industrial cities the world over and those of modern Western civilisation, and more recently Wheatley (1967) has emphasised how ethno-centre has been much of our thinking about the genesis of urban settlement.

Study of urban phenomena in the Indian sub-continent has

* Reprinted from *Geographische Zeitschrift*, Vol. 57, 1969, by permission of the authors and editors.

certainly suffered from unreality by forcing description into frames that do not apply and by carrying out exercises quite inappropriate to Indian urbanism. Even more stultifying to the advancement of urban geography in India than the falsities introduced by such fairly obvious distortions in an effort to apply Western models is the serious limitation upon descriptive analysis imposed by selection of those features and problems of Indian urbanism where analogies from study of Western cities do apply to the neglect of those either not represented in or not characteristic of the latter, the very ones indeed that make the Indian city distinctive. Among such features some are w despread and common to cities throughout the sub-continent. Others, however, may be more restricted in their geographical and cultural range, expressing rather a provincial or regional tradition and serving therefore as a basis for recognition of sub-types.

It is against the background of this situation that a preliminary analysis of the structure of the Indian city, examining its constituent elements as they occur together in distinctive association, is here offered. It follows and accepts the illuminating characterisations put forward by Spate (1950), Brush (1962) and Malik (1965), but is especially based upon the now fairly numerous studies of individual cities by Indian and Pakistan geographers. Much of this work, however, is unpublished in Ph.D. theses.

In any approach to the classification of the forms of urbanism, the great families of cities which share a common ancestry and history of development and which belong geographically therefore to major culture areas of the world must be recognised. The Indian sub-continent is such a culture area, the field of development of that family of civilisations distinguished and named by Arnold Toynbee as the Indic (parent) and Hindu (affiliated), the latter passing as a "universal state" successively into the Mogul raj and the British raj. It must be emphasised that, although urban forms in this great field of city development date far back into pre-history, when Mohenjo-daro and Harappa flourished four thousand years ago but disappeared with their precocious civilisation in the Indus basin, and were re-introduced with the Hellenic intrusion of the fourth century B.C. which produced Taxila, there does not appear to be such a long continuous urban tradition. Thus, whereas these ancient cities were of formal lay-out in a regular grid, indigenous cities of later times show no such evidence of planned development. These cities, whether they have grown around or alongside a fortress, royal palace or holy place, or have emerged from a village, are alike irregular and amorphous. Where urban development was associated, as so often, with a Mogul fort of the sixteenth or seventeenth century the latter typically stood separate from the walled city.

Attention has been drawn elsewhere (Smailes 1953) to the special significance of towns as expressions of imperial expansion and the intrusion of conquerors, who seek security in urban concentration and whose urban character also reflects their roles as administrators, traders, or military pensioners in contradistinction to an indigenous peasantry. Thus two of the commonest types of gathering point for urban development, the fortress and the trading post, are very typical features of such an impact of cultures. India in the successive phases of the Mogul raj and the British raj provides a classic illustration and its cities clearly reflect in their composite and disjunctive structure the plural society of imperialism, together with its modification in the aftermath of the past twenty years during which urban growth has proceeded with accelerated pace.

To a measure of duality already imparted to inland cities by the Mogul conquest, the European penetration of India from new trading posts on the coast added an even more distinct dichotomy in some Indian cities during the eighteenth and early nineteenth centuries, even before the building of railways broke down the earlier degree of isolation and facilitated the spread and strengthening of imperial administration. The latter was largely effected through the medium of cities, which extended their grip upon the rural countryside both as centres of trade and of administration. In the pre-railway age, however, the duality of indigenous city and European colony, although established, was limited in its representation and development.

In that short period of hardly a century, between the 1850s and the Independence of 1947, which may appropriately be called "railway British", the distinctive dual city form was diffused over the sub-continent and developed its full range of distinctive characteristics. Before we consider the later modifications and additions it has experienced, it is appropriate first to examine the Indian city at its fullest expression of the dual society of the

TABLE 3.1 : The Indian city : Growth structure

	Urban Duality	
	Indian elements	European elements
Pre-urban nucleus	Fort (esp. Muslim), e.g. Agra, Golconda, palaces ("garhis"), temple (esp. South India), village	Fortified European trading post, e.g. Madras, Calcutta
		Modern industrial works, e.g. Jamshedpur
1550 Mogul conquest and early European penetration	Old City Mogul city foundations and modifications of older cities with strengthened fortifications, Muslim quarters ("mohallas") and institutions, e.g. mosques	European annexes Colonisation by trading posts and military garrisons coastal (chiefly after 1600) and inland (after 1750)—establishment of segregated additions
		(a) Cantonment (military) (b) Civil Lines
	Accretion and bazaar ("chauk") development—but social differentiation by community groups rather than morphological by land use	Introduction of govt. institutions and other "British" services, including educational
		Clear-cut morphological units, with distinct functions
1850		As above continued

Railway development, from 1853 British Imperial Government, from 1858	Expansion by commercial and industrial development—colonisation by business premises ; annexes of modern industrial works and "bustis"	Railway stations and Railway colonies
1947	Coalescence of built-up areas by swelling and infilling	
Independent India Accelerated industrialisation	Addition of industrial estates, planned housing colonies (for former slum dwellers, industrial immigrants, and refugees), and unplanned "bustis".	
Gradual extension of motorable roads and modern road transport	Indian colonisation of former "British" quarters by officials and higher income-groups	
	Developing morphological differentiation by function, and of residence types expressing income status	

Note : In many Indian cities the major morphological periods represented are pre-railway, railway British, and post-Independence.

TABLE 3.2 : **Indian city morphology**

Morphological element	Functions	Lay-out	Fabric, building types and materials	Relief
Old walled city	Mixed : residence (incl. high-status Indian), handicrafts, trade	Irregular, congested	Traditional styles, pucca (often dilapidated) : some replacement by concrete	3 to 5 storeys temple and mosque salients
	Juxtaposition and superposition			
	Ethnic groups in quarters ("mohallas"), occupational streets ("paras") and alleys			
"Modern" extra-mural extension	Mixed : residence, handicrafts, with some factories, neighbourhood bazaars, and wholesale markets	Irregular, congested, but more spacious towards outskirts	Mainly pucca (brick) Kacha on lowlying ground along streams and in ravines	1 or 2 storeys but chawls higher
		Factory ribbons along approach roads	Engulfed villages of kacha housing "Bustis"	Few salients except factory chimneys
		Some open spaces (tank sites and graveyards)		

Cantonment	Military garrison	Regular grid, with liberal open space	European bungalows in gardens	Single storey, punctuated by Church spire and occasional large institutional building
	High or middle-class residence		Institutional buildings	
Civil Lines	Civil administration	Less regular plan, but spacious, with wide, tree-lined roads, gardens, and other open space	Institutional buildings and "Western" business premises	Single storey but some salient public buildings
	"Western" services, High or middle-class residence		Architectural styles more mixed than in Cantonment pucca, incl. small dwellings	Single storey
Railway Colony	Railway depots and employees' dwellings	Very regular		
Post-1947 estates and colonies	Factory industry and residence (according to income group), with neighbourhood services	Open, planned but not always rectilineal	Brick or concrete Stereotyped govt. or company housing in very small units	Low, 1 or 2 storeys
	Squatters' colonies ("bustis", "Jhuggis" or "cheris")	Irregular	Kacha	

British raj—the indigenous city and alongside, but physically distinct, the British annexes.

The descriptive analysis that follows has also been summarised in Tables 3.1 and 3.2. One shows the growth structure of the Indian city as it has developed over successive morphological periods, the other associates features of function, lay-out, fabric, and urban relief that characterise the components. Further, a generalised diagram (Fig. 3.1) depicts the arrangement of various elements as they group together in a characteristic geographical association.

In sharp contrast to the formalism, often severe, of the street pattern of the Western components nearby, the indigenous city is formless. Its structure everywhere is amorphous, except in the few cases where indigenous people occupy a city that is entirely modern and laid out under European influence, as in the untrammelled sprawl alongside the fortified trading posts on the coast, such as Calcutta and Bombay, or in the newly established urban centres of the modern canal colonies of the Punjab, such as Lyallpur. The city of Jaipur, laid out for its princely ruler in 1727 with spacious rectilineal streets, is another exception, but one which may derive, however, from contemporaneous planned cities in Europe.

In the Indian city of pre-European origin lack of plan is typical alike of the congested area within the perimeter of old walls and of the sometimes rather less crowded extensions outside, threaded by what were formerly country roads approaching the old city. Narrow, winding and congested streets, with jutting booths that encroach upon the highway, are so difficult for vehicular traffic to negotiate that thoroughfare is a misnomer for them, and the crooked side alleys, often cul-de-sacs, give only pedestrian access to the dense masses of buildings that fill in the skeletal pattern. Few streets have paved or metalled surfaces or are provided with culverted drainage ; more usually they are unmade, simply trodden earth, grooved by pestilential open sewers and encumbered with garbage.

Although it is often dilapidated and crumbling, the old property, built in traditional styles, consists of substantial buildings of stone blocks or sun-dried bricks (*kakaiya*), faced often with plaster and roofed with curved tiles. Different trades and crafts congregate in particular streets or lanes, as they did in the

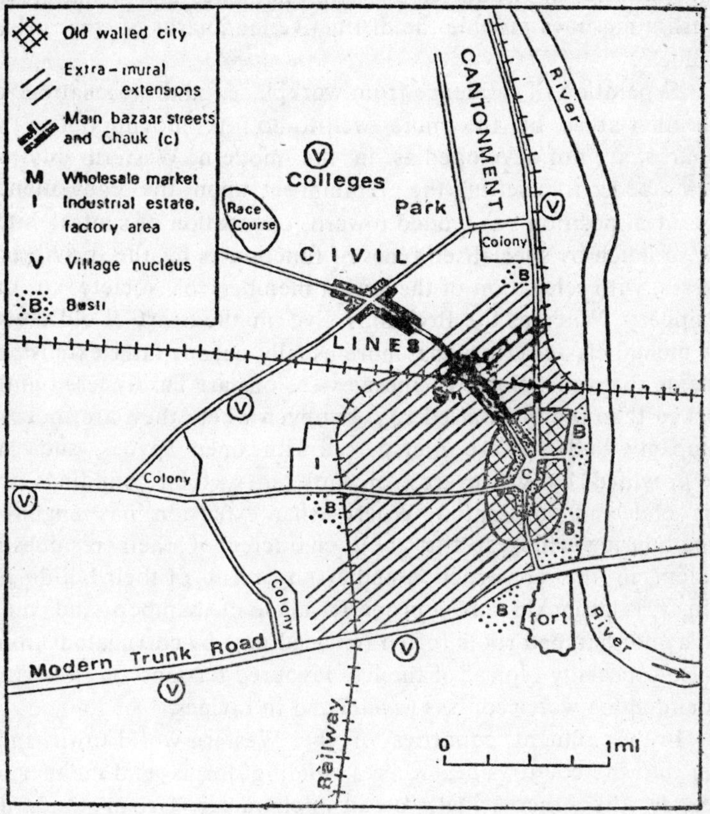

FIG. 3.1 : *Indian City* : *Generalised diagram*

medieval cities of Europe, but the several functions of residence, trade, storage, and craft industry are intermingled in the closest association. The trader or craftsman lives above or beside his shop or workshop and, although the main food market (*chauk* or *sadar*) occupies a relatively convenient central position at the main convergence of streets it shares retail distribution with other bazaars by the gates or dispersed among the neighbourhoods they serve. The essential structure of the city derives from its occupation by distinct ethnic and social groups and is not expressed in terms of differentiated land use. Each neighbourhood has its own set of institutions, in particular place of worship.

The pinnacles of Hindu temples and the domes and minarets of Muslim mosques provide the distinctive landmarks of the urban profile.

Separation of residence from workplace, and evacuation of the inner areas by the more well-to-do folk, moving out to the suburbs, are not developed as in the modern Western city. In such a pedestrian setting the premium set upon the convenience of central position has tended towards occupation of central sites not so much by specialised types of function as by the privileged classes, with relegation of the lowest members of society to the periphery. Whereas the Brahmins live in the heart of old cities, the menial classes (Harijans) more usually occupy later extensions outside the walls. Some of these extensions are hardly less tightly packed than the old walled city, but even where they are not so compactly built and are interspersed with open spaces, such as are provided by graveyards or tank sites, their buildings are less solid and substantial. Where urban extension has engulfed neighbouring villages, the *rural* character of their residues is evident in the construction and arrangement of their buildings. Flimsy, ramshackle *kutcha* property, made of bamboo and mud walls and thatched roofs (often replaced now by corrugated iron) is also especially typical of the less favoured terrains on lowlying ground along water courses (*nalas*) and in ravines.

In the affluent countries of the Western world towns spill over into the countryside, where building forms and manner of life are alike increasingly urban in character. Urban standards prevail, and both the buildings and their inhabitants become less and less differentiated. But in India it is the rural countryside that seeps into the cities, which swarm with villagers keeping their traditional mores as best they can. Rural building forms intrude into the townscape, farm animals wander about the thronged streets, and much of the essentially outdoor domestic life of the village persists among the new-comers.

In some Indian cities, however, notably Bombay, Calcutta and Kanpur, factory industry was already a powerful agent of urban growth before Independence and, in the nineteenth and early twentieth centuries, the spread of the urban area had taken place in the form of industrial tracts where large factory buildings such as textile mills, with tall chimneys, punctuated drab masses of slum dwellings, cheaply built of bricks. Whether low *ahatas* or *katras*, in rows within compounds, or multi-storeyed *chawls*, these tenements were appallingly overcrowded and occupied in shifts for sleeping.

Besides modern large-scale factory industry, which settled outside the already crowded old part of the city, some wholesale markets sought space in similar locations. Sometimes, too, old palaces and summer residences of the wealthy, together with compounds where menials lived, had been established outside the old city but were enveloped by building extension. Straggling ribbons of modern building may stretch some distance along the country roads into the surrounding farmland, and this is a specially noteworthy feature of factory development in recent decades, but the compactness of the Indian city and its sharp definition against the countryside are far more persistent in a country of still undeveloped motor transport than in the West.

Very different in appearance, as indeed in almost all respects, from the indigenous city, and standing physically distinct and apart from it, were the annexes planted by the imperial power for its expatriates and agents. The functionally distinct elements of military garrison (Cantonment), civil administration (Civil Lines), and railway colony, making their appearance in this sequence, were all initially set apart from the indigenous city and separated from it by a *cordon sanitaire*. Compared with the native city they were all, though in varying degree, organised and regular, spacious and sanitary, with buildings far more clearly distinguished and grouped according to function. Commercial, administrative, and other institutional buildings are all clearly differentiated from dwellings and the latter, while all of solid "pucca" construction, range from the opulent detached "kothis" of the highest officials to the stereotyped terraces of little boxes occupied by lower-grade railway employees. Furthermore, they are finely sorted, not as in the old city where mansion and hovel may lie cheek by jowl.

First among the British elements to make its appearance in inland cities, the Cantonment or garrison quarter displayed its special function by the presence of barracks, ordnance depots, drill ground and rifle range, together with officers' residences in the form of large bungalows with verandahs, set in spacious and colourful gardens along broad paved roads laid out in a formal grid. Some accompanying service features were also present in the Cantonment, such as a cluster of local shops (the Military Bazaar), church, and polo ground.

In some of the space between the Cantonment and the old city, or alongside the former, there developed the Civil Lines, less rigorously planned and systematically laid out than the Cantonment, but like it in the spaciousness of its tree-lined road system and in the liberal provision of private gardens surrounding bungalows, as well as other open spaces, including parks and racecourse. Here the personnel of civil administration and colonial activity were accommodated with the institutional equipment their presence and way of life called for. Thus along the main street or Mall and its offshoots are public buildings such as law courts, tax offices, post office and police station, as well as British-style shops, offices, and banks, together with hotels, bars, cafes, and cinema. Elsewhere in this locality are the bungalows where the British community of officials, commercial managers, and missionaries formerly lived. *Western* services naturally located in the Civil Lines also included churches, hospitals, and schools and colleges that were amply provided with grounds. The educational institutions were sometimes sufficiently congregated and extensive to form a well-developed precinct, which may have expanded into a University area. It was the usual antecedent of the University in the Indian sub-continent.

In the later part of the nineteenth century the railway colony added yet another separate component to this assemblage. Besides maintenance shops and marshalling yards, which occupy an extensive area, the colony includes housing for railway employees laid out on a very uniform grid in types graded according to the work status and income of different categories of employees. The railway colony was the meanest, most drab and regimented element among the Western components, its dreary tone sometimes emphasised by the use of uniform paint throughout, but at least it offered certain minimum standards of sanitation and residential amenity, both in its streets and buildings, that were quite foreign to the neighbouring indigenous city.

To the major periods of "pre-railway" and "railway British" that stand out in the development of the Indian city in this dual form the past twenty years of Independence have added another during which some modification of the structure described above has taken place, while at the same time large-scale accretion has introduced some new forms and emphasised others that had already made their appearance earlier.

Even before 1947 the distinction between the military and civilian annexes had often become blurred, and the Cantonment and Civil Lines tended to merge as the Western component and community by the side of the native city. Since then, as British officials and other expatriates have moved out, these localities have been colonised by their successors, the officials of the new regime and the more Westernised and wealthier commercial and professional Indians or Pakistanis. Cement cottages have also been added to the Civil Lines for housing the lower professional classes and minor officials. Inevitably the onetime cleavage between two civilisations, which was so clearly expressed in urban duality, has given way to the social gradation of a native society that is far from homogeneous but not simply dichotomous, and the physical separation by a *cordon sanitaire* has at the same time become less distinct or has disappeared as buildings have encroached upon the intervening space.

Nevertheless much of the massive extension and infilling of urban areas in India since 1947 is easily recognisable in its forms as pertaining to a new morphological period. Spread of the built-up area has generally made a far greater contribution than has physical re-organisation of urban space, but increasing replacement of old buildings in the hearts of cities by specialised commercial premises built in concrete, and elsewhere in the indigenous city by blocks of tenement housing has taken place.

Recent acceleration of the process of urbanisation has seen the emergence of many more places above the threshold of townhood by any definition, but it has also brought about great extension of the established and larger cities. The new urban dwellers are country folk drawn from the countryside by the attraction of urban employment both in new and growing industries and in developing central services. Especially in the years of painful adjustment after Partition mass movements of refugees greatly swelled some urban populations on both sides of the new frontiers.

Modern industry is often accommodated in large units, or in factory estates, located on the outskirts or in the larger interstices of the built-up area, invariably by the railways, for large scale industry in India is still so tied. In such locations large factories find space to spread and factories are grouped together. Housing for their workers has been provided in separate, though of course accessibly situated, housing estates, built either by the government or by the industrial companies. The dwellings tend to be monotonous and unprepossessing, but in certain essentials these housing estates are far more adequate and sanitary than earlier workers' housing. They derive from planning concepts and ideals borrowed from the West, though often debased and cheapened in the process of application by the sheer exigencies of the local situation. Naturally they show the use of new mass-produced materials such as concrete, whereas brick prevailed earlier.

Such has been the rate of growth of urban population, inflated often by mass influx of refugees, that housing provision has lagged seriously behind. Large numbers of incomers have perforce provided their own makeshift homes of cheap and waste materials, such as petrol-cans and sacking, in squatters' colonies (*bustis*) on any convenient open spaces within or on the outskirts of the urban area. By the very nature of their development the bustis are formless, amoeba-like tracts of slums, urban in location and density, but rural in their lack of utilities and services even of the most elementary sort and in the out-door mode of life except during the monsoon rains. By reason of this last feature the lot of their inhabitants, however indigent, is perhaps preferable to that of miserable occupants of chronically overcrowded rooms in the inner areas of Indian cities.

Such *bustis* (*jhuggis* in West Pakistan, or *cheris* as they are known in southern India) house a very considerable proportion of the sub-continent's urban dwellers at the present time. The overall proportion is perhaps 10 or 15 per cent, but it is much more in some northern cities. They demonstrate the magnitude of a housing problem that is all too characteristic of contemporary urbanisation in underdeveloped countries and in comparison with which the housing problems of the affluent societies pale into insignificance.

There can be little doubt that the pace of urbanisation in India has outstripped that of industrialisation. Beyond the degree of provision of productive employment in cities, the overpopulated countryside has shed some of its surplus population to swollen cities where the rural poverty is shared through excessive employment in service occupations. Far more persons are present and supported in a fashion in the Indian city than its economic base warrants, and the total service function is comminuted among a swarm of underemployed tertiary "workers". Shops and street-traders are far more numerous than rational organisation of retail distribution would justify, there are too many office employees to make possible efficient administration of commerce or government and if the ubiquitous rickshaws and cycles may often seem to achieve the most effective means of movement within the Indian city this is because of the dearth of vehicular thoroughfares in the Western sense. From the teeming countryside villagers pour into the cities in numbers in excess of the expansion of industrial employment, and eke out an existence there in overmanned trading and other services.

On the other hand, especially in the metropolitan cities, the accommodation of the most highly centralised functions of government and business has fostered the development of concentrations of office buildings in Western-style premises at the expense of the mixed functions and traditional buildings that were there before. Yet the single dominant commercial core or Central Business District of the modern Western city has no true counterpart in Indian city structure. Not only is this because central functions have often been divided between indigenous city and the Civil Lines of British establishment. The essential character of Western "City centre" or "downtown" derives from geographical concentration of centralised services at a nodal, accessible point and presupposes a commensurate development of transport services. In Western cities today, because of increasingly acute problems of traffic congestion in the centre, the dominance of the Central Business District is in some respects, especially for shopping, being reduced. In the absence of easy accessibility from a wide range in the Indian city, it has never been possible for it to be asserted to the degree that had become characteristic in the West. The immobility that goes with poverty and much less developed urban transport systems makes far greater reliance upon local neighbourhood bazaars as compared with the main bazaar or "chauk" and centrally situated shops. In its economic as well as in its social aspects, the life of the Indian city is far more dispersed. In Western cities the most close-knit social neighbourhoods have developed in association with poverty and/or social ostracism. In the cities of the Indian sub-continent in extreme degree widespread poverty has combined with social and ethnic barriers to promote a cellular structure.

Nevertheless, as the British expatriates have been replaced in the distinctive quarters they had created by the upper echelons of the indigenous society, and as the most recent urban extension takes the form of housing of different types in compact or fairly compact groups that are closely geared to the means of their occupants, a Western-style differentiation of the whole urban area in terms of income-status is coming to the fore. As social divisions according to birth slowly give way to those directly related to achievement as reflected in wealth, the changing structure of the new national societies both accommodates itself to the urban geography inherited from the dual society and morphological periods of the past and expresses itself directly in the most recent accretions to this older townscape.

If the urban duality of the imperial period is now outdated, a legacy from the past in the contemporary Indian city that is bound to fade with the passage of time, it may be asked whether in the qualitative expressions *pucca* and *kacha*, as applied to distinguish buildings and roads, there is a basis for making a continuing two-fold distinction of more contemporary significance in the urban scene. Does it warrant recognition as one that ought to be established in all morphological survey and mapping of Indian cities by geographers or does the connotation of these terms vary so in time and place that they are too loose in application ?

In putting forward this descriptive model of Indian city structure and venturing generalisations, it must be mentioned that there are gaps in the existing knowledge, or at least in such as is published and thereby disseminated. How complete and perceptive in essentials of portrayal is the picture that has here been presented ? What sub-types might be elaborated, by what criteria, and with what distribution-patterns ? Especially, how limited in number and range is the Indian city as described here ? Obviously, even if the generalisations drawn here truly epitomise a type, it is represented only in a particular upper range of the urban hierarchy in India. How narrow ? How far down from the top of the hierarchy, below the two greatest metropolitan cities that are entirely creations of the British period, does this characterisation apply ? The constituent elements that express the dualism derived from the British raj clearly are limited to those urban centres that were already sufficiently developed in the period before 1947 to have garrisons, communities of the Indian Civil Service and other British expatriates, and railway stations. There were many others that already merited recognition as towns, whether by reason of their size, functions, or urban equipment, and their number goes on increasing. Indeed, by the somewhat imprecise criteria of the Census definitions there are more than three thousand towns in the subcontinent where, although rural life is still the rule, there must nevertheless be well over one hundred million town dwellers.

As urbanisation proceeds with gathering momentum, more and more settlements are crossing the threshold of townhood. In some the mainspring of growth is industrial development, but in many more it is concentration of service functions as centralised institutions are provided at strategic nodal centres for the rural population of the surrounding districts. The attributes of these emergent towns in the lower ranks of the urban hierarchy and differences among them, whether in size, functional emphasis, or cultural associations deserve close study.

Finally, should one recognise in the ensemble of features described above, embodying as they do certain constituent elements unrepresented in places lower down the hierarchy, the basis of a distinction of such magnitude as to merit recognition in the context of the Indian sub-continent as that between "city" and "town" ? Should one be justified in reserving for the type described above the appellation "Indian city" ? If so, characterisation of the essential features of the Indian town and its variant types is awaited.

FOUR

BRIAN J.L. BERRY AND PHILIP H. REES

THE FACTORIAL ECOLOGY OF CALCUTTA*

THIS paper represents a modest attempt to extend cross-cultural research in urban ecology, using a factorial study of the city of Calcutta as an instrument for drawing together diverse but related streams of thought. The extension is but a prologue, however, for systematically executed comparative ecological analysis remains an unattained goal.

A VARIETY OF INSIGHTS

Relevant insights and intuitive analogies are many. For example, in his mid-nineteenth century classic, Kohl (1841) devoted an entire chapter to the internal structure of cities. Urban structure, he said, might be viewed vertically as a series of layers as in Figure 4.1 : (1) the ground floor (Rez-de-Chausee) contains the establishments and living quarters of the businessmen ; (2) the first floor (Bel Etage) is the "area of wealth and pleasure", the seat of the nobility ; whereas (3-5) "upwards" people with lower income reside ; the same (a and b) applies to the subterranean levels. These layers, he continued, translate into "arches" of homogeneous social structure, so that the economic and social importance of the population declines as you move out from the centre of the city. However, some outer rings might contain a wealthy population because of "a growing sympathy for the

* Reprinted from *American Journal of Sociology*, 1974, No. 5 (March 1969), pp. 445-491 by permission of the authors and editors.

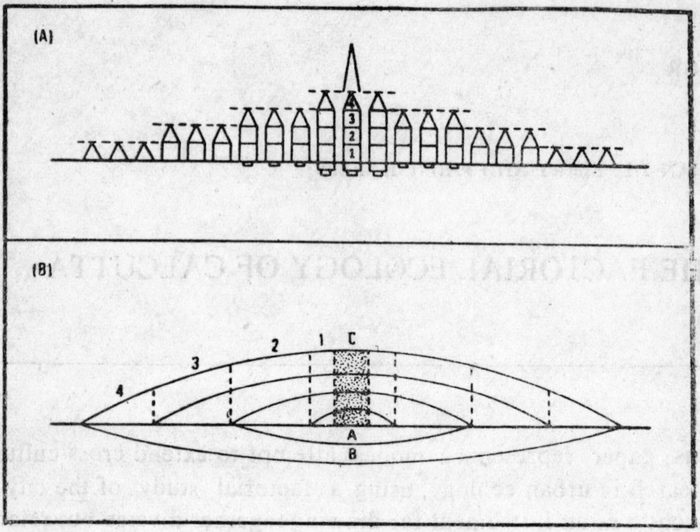

FIG. 4.1 : *Kohl's concept of social space (after Peucker)*

countryside".*

Parallels with this nineteenth-century European view can be found in studies of Indian cities. For example, Chatterjee (1960) concludes, in looking at the ecology of Calcutta's neighbour, Howrah :

> The influence of the caste system is reflected in the usual concentration of the higher castes in the central areas of good residential localities, while the lower caste groups usually occupy the fringe . . . The people still attach more importance to these centrally-situated residential areas . . . . Thus, in spite of the modern development of road transport, the residential decentralization or movement towards the fringe outside the old residential areas is not very marked.

Sjoberg (1960) generalizes the pattern to all "pre-industrial" cities and contrasts it with "industrial" urbanization :

The feudal city's land use configuration is in many ways the

* The authors are indebted to Thomas G. Peucker for this example.

reverse of that in the highly industrialized communities. The latter's advanced technology fosters, and is in turn furthered by, a high degree of social and spatial mobility that is inimical to any rigid social structure assigning persons, socially and ecologically, to special niches. . . .

(There are) three patterns of land use wherein the pre-industrial city contrasts sharply with the industrial type : (a) the pre-eminence of the "central" area over the periphery, especially as portrayed in the distribution of social classes, (b) certain finer spatial differences according to ethnic, occupational, and family ties, and (c) the low incidence of functional differentiation in other land use patterns.

According to the classical papers of the Chicago School (Park 1936), the ecology of the industrial city—Sjoberg's norm—stems from a process of competition between various population or interest groups in the city, the dominance of particular groups within the natural areas of the metropolitan community (Zorbaugh 1926), with change taking place by the invasion of a natural area by a competing group, and the ultimate succession to dominance of that group in the area (Rees 1969).

The process, the argument continues, derives energy from city growth, such that older population groups move from their original home to the periphery as their economic status improves, to be replaced at the city centre by new arrivals. It results in Burgess' concentric zones of successively increasing status, subdivided into small natural areas on the basis of race "the Black Belt", or ethnicity "Little Sicily", or type of structure "Residential Hotels", as depicted in Figure 4.2 (Park, Burgess and McKenzie 1925). It is this zonal result that Sjoberg contrasts with the pre-industrial form.

There have, of course, been alternative views of the spatial patterns or the determinants of the internal structure of the twentieth-century industrial city. Hoyt (1933), for example, argued that high-rent and low-rent neighbourhoods occupied distinct sub-areas of the city and that these were not aligned concentrically about the city centre but were distributed in a sectoral fashion. The spatial pattern of the city's rental areas was determined in Hoyt's view by the choice of those who could afford the highest rents. They pre-empt land along "the best existing transportation lines", "high ground—free from the risk of

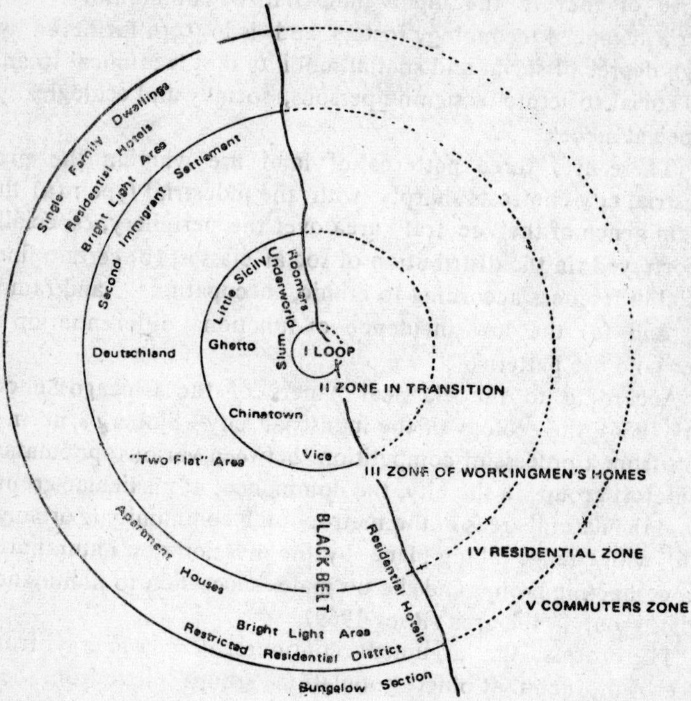

FIG. 4.2 : *Burgess' zones and ethnic communities*

floods", and land along lake, bay, river, and ocean fronts where such water fronts are not used for industry. Others have suggested the significance of persistent cultural values attached to areas (Firey 1947) and of "multiple nuclei". Yet others (Harris and Ullman 1945) have tried to identify and characterize new "post-industrial" ecologies characterizing contemporary metropolitan growth (Friedmann and Miller 1965).

BOSE'S ANALYSIS OF CALCUTTA'S ETHNIC DIVERSITY

The most important ecological study of Calcutta available to this time, by Bose (1965, 1968), is phrased largely in terms of ethnic variability and cultural differences. Bose depicts Calcutta as "the scene of a major confrontation between the enduring institutions of old India—her caste communities and diversity of ethnic heritages—and the pressures and values arising from the process of urbanization that pressages India's industrial revolution."

Traditionally, Indian society was based upon caste : Brahmin priests and teachers ; Kshatriya warriors and rulers ; Vaisya land-owners, merchants, and money-lenders ; Sudra cultivators, workers, and artisans ; and the "untouchables" (today's "scheduled" castes) were the classical ideals in a complex system of more than 12,000 caste communities. Each was identified with occupation and maintained by intra-marriage and spatial monopoly of particular trades in particular locales. Ranking of occupations into a hierarchy which generally coincided with ritual practices and residence of different castes in separate quarters maintained distinctiveness in the system and provided social content for the urban ecology.

It is this pattern that is being modified in Calcutta, India's largest city, according to Bose. The city of Calcutta is a legal unit of more than three million people set within a metropolitan area of seven million (according to the 1961 census). Created by the British during the eighteenth century, and once the seat of the British raj, it is today a leading industrial centre, a major port, a national metropolis, and the focus for eastern India (Berry 1966, Ghosh 1961, Mitra 1963).

Bose points out that Calcutta still houses many traditional communities for, although the city was a British creation, it was occupied by Indians who brought to it traditional Indian institutions.

Bose concludes :

> The map of Calcutta shows a highly differentiated texture. Ethnic groups tend to cluster together in their own quarters. They are distinguished from one another not only by language and culture but also by broad differences in the way they make their living. Naturally there is a considerable amount of overlap but this does not obscure the fact that each ethnic group tends to pursue a particular range of occupations.
> It can be said, therefore, that the diverse ethnic groups in the population of the city have come to bear the same relation to one another as do the castes in India as a whole. They do not enjoy monopoly of occupation, as under caste, nor are they tied to one another by tradition in reciprocal

exchange of goods and services. There is also no ritual grading of occupation into high and low. But preference for or avoidance of some kinds of work are expressed in class differences among occupations, as can be observed elsewhere in the world. The social order of Calcutta might, therefore, seem to be evolving through a transitional stage, in which caste is being replaced by an increasingly distinct class system.

Actually, the superstructure that coheres the castes under the old order seems instead to be reestablishing itself in a new form. Calcutta today is far from being a melting pot on the model of cities in the U.S. There the Irish, Italian and eastern European immigrants have merged their identities within a few generations. The communal isolation of the first generations was quickly reduced by occupational mobility in the expanding American economy and by the uniform system of public education that Americanized their children. In Calcutta, the economy is an economy of scarcity. Because there are not enough jobs to go around everyone clings as closely as possible to the occupation with which his ethnic group is identified and relies for economic support on those who speak his language, on his coreligionists, on members of his own caste and on fellow immigrants from the village or district from which he has come. By a backwash, reliance on earlier modes of group identification reinforces and perpetuates difference between ethnic groups.

AN APPROACH VIA FACTORIAL ECOLOGY

If Calcutta is, as Bose argues, the scene of confrontation between traditional Indian institutions and industrial urbanization, then one should be able to see in its ecology not simply the ethnic diversity of region, caste, and culture that he describes, but also the two further ecological themes of the pre-industrial city categorized by Sjoberg (pre-eminence of centre over periphery, low incidence of functional differentiation in land use) and an increasing admixture of the ecological patterns of the industrial city. One way to determine whether this is the case is by assuming the analytic stance of factorial ecology.

Factorial ecology is an outgrowth of social area analysis.

Since there exists semantic confusion about this term, some clarification is required (Table 4.1).

TABLE 4.1 : **Typology of social area analysis**

Social Area Analysis *Sensu Lato*	Method employed
TYPE 1 : Social area analysis *Sensu Stricto*	Construction of Shevky-Bell indices
TYPE 2 : Factor analysis of social area *Sensu Stricto* variables	Factor analysis of Shevky-Bell index variables
TYPE 3 : Factorial Ecology	Factor analysis of Guider set of socio-economic variables, including the Shevky-Bell set

Social area analysis, sensu stricto

The term "social area analysis, sensu stricto" applies only to that mode of analysis originally outlined by Shevky, Williams, and Bell in their studies of Los Angeles and San Francisco (Shevky and Williams 1949, Bell 1953, Shevky and Bell 1955). From a number of postulates they derived three basic constructs which they considered describe the way in which urban populations are differentiated in industrial society. The three constructs were called social rank (economic status), urbanization (family status), and segregation (ethnic status).* They then proposed three indexes, one per construct, each made up of from one to three census variables, designed to measure the position of census tract populations on scales of economic, family, and ethnic status. The analysis also made possible the classification of census tracts into social areas based upon their scores on the indexes.

Social area analysis, sensu stricto, has been criticized both on theoretical grounds (the theory underlying the constructs) and for empirical reasons (the method of dimensioning the constructs) (Hawley and Duncan 1957, Duncan 1955). Bell (1955) attempted to meet the empirical objections that the social area analysts selected measures on the assumption that the constructs were correct but failed to provide a test of their validity. He used

* The terms in parentheses are Bell's ; the preceding terms are Shevky's.

factor analysis to show that, in the cases of Los Angeles and San Francisco, the census measures selected to construct the indexes in fact formed a structure consistent with Shevky's formulations. Van Arsdol, Camilleri, and Schmid (1958) went on to extend Bell's test of the Shevky model to some ten cities, six of which confirmed the validity of the Shevky indexes as measures of his constructs. The fact that four cities did not conform to Shevky's constructs suggested, however, that the existence of the constructs should be left as an empirical question to be determined by the patterns in the variables rather than one to be assumed correct, *a priori*. The logical extension of this argument is that many more variables detailing the way in which census tract populations vary according to socio-economic characteristics should be included in any study and that factor analysis should be used to isolate the fundamental patterns of variation in the data, be they Shevky-Bell construct patterns or otherwise. Whether this viewpoint is valid may be debatable, but nonetheless it constitutes the premise on which the balance of this paper is founded.

Factorial ecology

Factorial ecology is the term now used to characterize studies involving this application of factor analysis to ecological study. A data matrix is analyzed containing measurements on m variables for each of n units of observation (census tracts, wards . . .), with the intent of (i) identifying and summarizing the common patterns of variability of the m variables in a smaller number of independent dimensions, r, that additively reproduce this common variance ; and (ii) examining the patterns of scores of each of the n observational units on each of the r dimensions. The dimensions isolated are an objective outcome of the analysis. Interpretation of the dimensions (factors) depends on the nature of the variables used in the analysis and the body of concept or theory that is brought to bear. Theory provides the investigator with a set of expectations regarding the factor structure which can be compared to the actual set of factors produced. This comparison was made formally by Van Arsdol, Camilleri, and Schmid but somewhat less formally by later writers. What is impressive, however, is that studies of American cities have, by and large, succeeded in isolating the three social area dimensions originally proposed by Shevky : socio-economic status, family status, and ethnic status.

Moreover, it becomes increasingly evident that each of these dimensions captures the essential features of one of the classic spatial models (socio-economic status—Hoyt, family status—Burgess, ethnic status—"segregation" studies, Firey), so that in Western metropolises born in the industrial age and populated by a variety of races or national groups, "the models are independent, additive contributors to the total socio-economic structuring of city neighbourhoods" (Berry 1965).

Those indexes that measure the socio-economic status of individuals or groups vary principally by sector ; those variables which measure the family and age characteristics of the population vary principally by concentric zone ; those measures that isolate a minority group within the population show a tendency for that group to cluster in a particular part of the metropolis. Such is the conclusion of Anderson and Egeland (1961) in their analysis of the spatial variance of a number of socio-economic measures within four medium sized American cities (Akron, Dayton, Indianapolis and Syracuse). Those variables that were thought to measure social status were combined into a "social rank" index, and those variables that were considered indicative of family status and degree of urban acculturation were combined into an index of "urbanization" in the manner originally developed by the social area analysts. An analysis of the variance of these two indexes by sector and by concentric zone revealed that social rank varied principally by sector and urbanization by concentric ring. A third dimension corresponding to their "segregation index" is also present and describes the concentration of particular minority groups in limited neighbourhoods of the city.

This basic triad of spatially arranged social dimensions can be superimposed to form, at the intersections of sectors, zones, and segregated areas, communities of similar social, family. and ethnic status. Figure 2.4 provides an idealized picture into which distortions can be successively introduced to closer approximate reality.

However, there is a further complication. The zones within the segregated area occupied by the minority group do not correspond to the general life cycle zones of the metropolis ; the segregated area is a microcosm of the whole, compressed spatially, reproducing in miniature the metropolitan-wide pattern. This modified pattern is then further distorted by city growth. "Tear faults" develop as zones cross sectoral boundaries with displacement of zones outward in the early-growth sectors. Finally, introduction of secondary work-place nodes—a heavy industrial area in the southern part of the city industrial satellites in a crescent—further changes the form of the model for the metropolitan region.

Behavioural basis : the residential location decision

The attraction of the classical ecological view of the industrial city is that it postulates a process of group competition and mobility that produces the spatial structure of the city's population. In Hoyt's model, on the other hand, the operations of the real estate market are not spelled out, though he does recognize the importance of the decisions of prestigious individuals in the location of the high grade rental sector.

An alternative view of the processes producing the structure may be proposed in the case of the North American city, however. This view involves the behaviour of individuals and institutions. The inhabitants of the city are faced with a fundamental decision : where to live. The principal determinants of such a housing choice are three in number—the price of the dwelling unit (either in rental or in purchase value terms) ; its type ; and its location, both within a neighbourhood environment and relative to place of work. These determinants have parallels in the attributes of the individual making the housing choice : the amount he is prepared to pay for housing, which depends on his income ; his housing needs, which depend on his marital status and family size, that is, his stage-in-life cycle ; his life-style preferences, which will affect the type of neighbour he will want ; and, finally, the location of his job. When the values of the two sets of characteristics match, a decision to purchase housing will be made.

The most important determinant is undoubtedly income. A large family with many children may need a large single family home, but unless family income is above that minimum threshold required by the mortgage company, purchase of this type of home will not be possible. Similarly, low wages may make it essential for a worker to live near his place of employment in order to minimize the costs of the journey to work. He may, therefore, be unable to satisfy his preferences as to type of neighbourhood.

It may be objected that other factors such as the nature of the school system in the community or the location of the prospective home relative to a golf course play an important role in the decision, but these influences become significant only if the individual's income is large enough for him to be able to ignore the more basic determinants. Gans (1967), in his participant-observer study of the inhabitants of a new suburban town in New Jersey, asked a sample of Levittowners what their principal reason was for moving to Levittown. Over 80 per cent gave house-related reasons for moving.

Gans's study of Levittown, New Jersey, contains many valuable illustrations of the way in which some of the residential choice determinants outlined above operate. In response to a perceived demand for middle-income homes, Levitt built houses with prices ranging from $ 11,500 to $ 14,500. To ensure that he got paid for the houses he built, Levitt told his salesmen to apply an informal lower income limit of $ 5,500 a year for the would-be house purchaser. Anyone with less income than that was not to be sold a house. In fact, a small percentage of house buyers did have lower incomes, but the majority had higher. The house price acted as a selection mechanism : the middle-range prices attracted predominantly lower-middle-class families with small minorities of working class and upper-middle-class families.

The houses were free-standing single family residences with three or four bedrooms, and, not unnaturally, young families with (or intending to have) children became their occupants. They were spacious and comfortable homes for a four- or five-member household, and the surrounding yard provided play space for the young children over whom the mother could keep an eye from the kitchen window. Since Levittown was an entirely new settlement, there were no neighbours in the community when the first inhabitants arrived, but the builder was careful to project a prestigious image in nearby Philadelphia (in order to counteract the unfavourable impression created by his other venture in the urban region which had experienced some trouble over racial integration and subsequent status decline), and, for the first two years, until legal action forced Levitt to integrate, discrimination was practiced against prospective Negro buyers.

The residential location decision was thus clearly a product of the interaction of the home purchaser and the home builder, a not unexpected conclusion ! Their choices either effect (in the case of a new community) or affect (in the case of an established community) the urban ecology, creating relatively homogeneous areas containing similar sorts of people (areas called "natural" in the earlier ecological literature and generally "social" today). The character of established social areas in turn influences subsequent housing choices of later neighbourhood occupants, because people seek compatible neighbours who share the same views on life (particularly on such important questions as the way in which children should be brought up) and find this compatibility by selecting their residence in the appropriate social area.

The choice of house and community type is only one part of the complete residential location decision, however. The housing consumer is also faced with the problem of locating his residence. This involves attitudes toward the journey to work, for the time and cost of commuting have to be traded off against the relative benefits of living in alternative communities that meet, within budgetary constraints, family needs. The lower the family or individual income, the more constrained will be the choice Thus, lower-status people live closer to their work than people of higher status.

Therefore, to summarize more generally the individual or family occupies a position, s_i in social space, determined by economic status and family status (Chombart de Lauwe 1952). The household matches this position with that of a dwelling located in an analogous position, h_i, in housing space, and of housing in a similar location, c_i, in an equivalent community space, whose axes comprise socio-economic status on the ordinate and familial characteristics on the abscissa. From a range of possible communities found in the same zone of community space, one dwelling in one community is selected on the basis of proximity to job location (if this is a constraint) or on the basis of other important neighbourhood characteristics, thus fixing the choices in real geographic or physical space.

An orderly social geography results through like individuals making like choices, through regularities in the operation of the land and housing markets, and through the collaboration of similar individuals in excluding those of dissimilar characteristics from their neighbourhood or in restricting certain minority groups to particular areas. The autonomous suburb is the prime example of the process of exclusion and the ghetto the most glaring illustration of the process of restriction. Sectoral patterning of such attributes of the neighbourhood residents as education, occupation, income, and of neighbourhood structural characteristics such as rent or value and quality of housing is a product of the differing abilities of various income groups to bear the costs of the journey to work. Lower-income workers, because of their restricted money budgets, must live close to their work (concentrated in the inner city around the central business district and along the rail and water routes radiating outward from it). The higher incomes of upper-status workers give them the freedom to locate their homes in areas of higher residential amenity—away from their places of work, away from the smoke and dirt of industry, and close to amenity features such as lakeshore and open space. The age structure of the population, average family size, and female labour force participation change as distance from the city centre increases : young families locate further from the centre than do older families. This pattern is a response to the change in house age and type as distance from the centre becomes greater—the houses are newer, and single family homes predominate as the city centre is left behind. It is the lower land values toward the urban periphery that make possible this land-voracious construction, and the increasing real income of home buyers makes possible their purchase. Finally, minority groups find themselves segregated from the rest of the population to a greater or lesser degree as a result of recent arrival in the city, discrimination in the housing market, or through choice of home in congenial communities.

COMPARATIVE EVIDENCE

Several possible factor structures

From these bases of residential choice, the ecological pattern of American cities is dimensioned cumulatively by socio-economic status, family status, and the constraints of race and ethnicity. What of non-American cities ? Do they exhibit the same patterns of variation in census tract populations ?

Pedersen's study of Copenhagen provides one of the most comprehensive urban ecological analyses to date (Pedersen 1967). From a matrix of fourteen socio-economic variable (related to age distribution, employment status, distribution by industry, household size, sex ratio, and female employment of the population) and seventy-six zones of the city, three basic factors emerged in both 1950 and 1960, namely : (i) an urbanization or family status factor, (ii) a socio-economic status factor, and (iii) a population growth and mobility factor. The first factor, when mapped and graphed, displayed the classic Burgess pattern of concentric rings ; the second factor exhibited sectoral distribution, with the exception that the central zones of the city were of uniformly low status ; the third factor exhibited highest values in the zone of new suburban development and in the low-status areas of the inner city. Pedersen devotes a separate chapter to detailed consideration of each of the three principal factors. Age structures of the communities of the city are considered as functions of the population growth and migration process. Income differentials in the suburban portion of the metropolitan region are shown to have narrowed over time as the area has been converted into residences for city dwellers from farmland, whereas the gap between inner city and suburban incomes has widened. Absent from Pedersen's findings is a differentiation of Copenhagen's population on linguistic or racial lines—such differentiation does not exist in Denmark.

Confirmation of such homogeneity of population in Scandinavian cities is found in Sweetser's comparison of Helsinki and Boston (Sweetser 1965b). In addition to the basic social area dimensions, the Boston study yielded a number of ethnic dimensions (Irish middle class ; Italian blue collar). The Swedish-speaking population of Helsinki, however, failed to load on a separate factor but rather loaded on the socio-economic status factor, since they were of generally higher status than the Finnish speaking population.

Studies of non-Western cities (Abu Lughod 1968b) highlight the cultural context of the American and Scandinavian factor structures and have led recent investigators to try to isolate those basic conditions in the urban system, social and spatial, which are necessary in order to produce the observed factor structures.

A principal conclusion of Abu Lughod's study of Cairo was that "no factorial separation between indicators of social rank and the indicators of family cycle stage could be obtained". This contrasts with the normal separation of these two sets of indicators in factor analyses of American city data matrixes. As a result, Abu-Lughod was led to outline, in an extremely effective way, the conditions that were necessary and sufficient to produce the dimensions of socio-economic status and family status that are found to have independent existence in almost all American cities conditions that were not fulfilled in the case of Cairo (Table 4.2).

These conditions shown in Table 4.2 in fact give rise to several alternative factor structures (Table 4.3) representing permutations of the three basic sets of variables—SES : socio-economic status set, LC : stage-in-life cycle of family status set, MG : minority group set,—only one minority group is discussed here, although clearly the argument could be extended to several. These combinations constitute the frame within which a factorial ecology of Calcutta can be evaluated.

CALCUTTA'S ECOLOGY : DIMENSIONS OF VARIATION, 1961

Principal axis factor analysis with normal varimax rotation of all factors with eigenvalues exceeding unity was used to analyze an 80-ward × 37-variable data matrix for Calcutta. The variables came from the 1961 census and represent all the quantitative materials available at the time we performed the analysis. They were obtained as special hand tabulations (as of the time of writing, the census tables have yet to be published for Calcutta), and they relate to family structure, literacy, type of employment, housing characteristics, and land use.

The units of observation were the city's eighty municipal wards—administrative units which, as Bopegamage (1966) points out, possess social, economic, and demographic homogeneity only by accident. This lack of homogeneity is regrettable, but perforce to be tolerated. Because, too, the wards vary in their total population size, this variable was left in the analysis as a "screen" for the possible influence of size differences. All other variables were converted to ratios (population density, average size of household) or to percentages (percentage of workers employed in

TABLE 4.2 : Conditions that produce dimensions of socio-economic status and family status

Factor conditions	Type of variables used	Necessary conditions
Socioeconomic status factor	Education, occupation, income	(1) "That the effective ranking system in a city be related to the operational definition of social status" ; (2) "that the ranking system in a city be manifested in residential segregation of persons of different rank at a scale capable of being identified by the areal units of observation used in the analysis"
Family status factor . . .	Family size, portions of the age pyramid, fertility	(1) "That family types vary, either due to 'natural' causes such as those associated with sequential stages in the family cycle, or to 'social' causes such as those associated with other divisions in society, whether ethnic, socio-economic or other" ; (2) "that subareas within the city are differentiated in their attractiveness to families of differing types" at a scale capable of being identified by the areal units of observation used in the analysis.
Disassociation between socio-economic status and family status dimensions . . .		Either (1) that there exists little or no association between social class and family type ; or (2) if there is some association between social class and family type, that (a) there is a clear distinction between stages in the family cycle, each stage being associated with a change of residence ; and that (b) "subareas within the city offer, at all economic levels, highly specialized housing accommodations especially

		suitable to families at particular points in their natural cycle of growth and decline" at a scale capable of being identified by the areal units of observation used in the analysis ; and (c) "cultural values permitting and favouring mobility to maximize housing efficiency, unencumbered by the 'unnatural' frictions of sentiment, local attachments, or restrictive regulations"
Separate minority group . . .	Proportion of the subarea population in the minority group, measures of the relative concentration of minority groups	(1) That the characteristic(s) differentiating the minority group from the rest of the population be of perceived significance in the social system, i.e., that the urban population be truly heterogeneous ; (2) that the minority group be residentially segregated from the rest of the population, at the scale of observation used in the analysis
Disassociation between minority group and socio-economic status dimensions . . .	Measures used above for minority group, measures of education, occupation, and income (and the distribution of resources)	(1) That the minority groups be residentially segregated from the rest of the population, at the scale of observation used in the analysis ; and either (2) that there exists little or no association between minority group status and socio-economic status ; or (3) that there is some association between social class and minority group status, that (a) the minority group still spans most of the social status range, though it may be concentrated at the lower end of the range ; (b) a fairly full range of housing accommodation quality be available for families within the residentially segregated area

TABLE 4.2 : (*Contd.*)

Disassociation between the minority group and stage-in-life-cycle dimensions . . .	Measures used above for the minority group, measures of age distribution, family size, and fertility	(1) That the minority group be residentially segregated from the rest of the population at the scale of observation used in the analysis ; and either (2) that there exists little or no association between minority group status and family status ; or (3) that if three is some association between family status and minority group status that (a) minority groups still span most of the family status range, though it may be concentrated at one end of the range, (b) a fairly full range of housing accommodation types be available for families within the residentially segregated area

TABLE 4.3 : **Factor structures**

Variable set	Highest loadings on factor
Combination 1 (Shevky-Bell)	
SES	1
LC	2
MG	3
Combination 2 (Cairo)	
SES, LG	1
MG	2
Combination 3 (American South, Montreal, Helsinki)	
SES, MG	1
LC	2
Combination 4 (Miami)	
SES	1
LC, MG	2
Combination 5 (Unidimensionability—Erewhon ?)	
SES, LC, MG	1
Combination 6 (Chicago)	
SES	1
LC	2
MG SES	3
Combination 7 (Toledo)	
SES	1
LC : Family Status	2
Urbanization	3
MG, SES	4

manufacturing and so forth).

The factor analysis routine utilized the University of Chicago's computation system and produced as output matrixes of descriptive statistics for each variable, simple correlations among variables, factor loadings for ten factors (correlations of variables with factors), eigenvalues indicating that the ten factors reproduced 79 per cent of the total variance or 59 per cent of the common variance, and orthonormal factor scores for each ward with respect to each factor.

Not all of these materials can be reproduced here ; instead, only the principal results of the analysis will be reviewed. However, the findings do reveal that alongside the rich ethnic variability described by Bose, Calcutta is also characterized by a broadly concentric pattern of familism, an axial arrangement of areas according to degree of literacy, and both substantial and increasing geographic specialization of areas in business and residential land uses, gradually replacing the former mixture of businesses and residences that were separated, rather, into occupational quarters. This mixture of pre-industrial and industrial ecologies thus lends support to the idea that the city is in some transitional developmental stage.

Factor 1 : A land use and familism gradient

According to factor 1, the larger peripheral wards of the greatest population size (factor loading 0.762) are also those with the greatest percentages of females over fifteen (loading 0.817) and married women (0.749), the greatest non-Bengali component (0.591), proportion of buildings solely dwellings (0.892), with a higher proportion of the houses mudwalled (0.545). Conversely, the smaller wards at the heart of the city have greater proportions of their buildings occupied by businesses and offices (−0.825) and by commercial establishments and restaurants (−0.518), greater proportions of the population employed (−0.796), a greater mixture of the "houseless" male population (−0.727), and a greater percentage of the population residing in institutions (−0.813).

A sixth-degree trend surface* fitted to the factor scores reproduced 78 per cent of the variance in the scores in terms of spatial regularity and shows a clear concentric pattern around the city centre (Fig. 4.3). There is thus a clear and strong land use and familism gradient in Calcutta from commercial core to residential periphery, modified only by the interjection of peripheral nodes at the cardinal extremities of the city. The commercial core houses most of the "sidewalk sleepers", largely young males, whereas the dominantly residential areas of the outer zones of the city have more balanced sex ratios and concentrations of mud-walled dwellings. Comparison of trend

* The equation is

$$
\begin{aligned}
F= & a_9+a_1X+a_2Y+a_3X^2+a_4Y^2 \\
& +a_5XY+a_6X^3+a_7Y^3+a_8X^2Y \\
& +a_9Y^2X+a_{10}X^4+a_{11}Y^4+a_{12}X^3Y \\
& +a_{13}Y^2X+a_{14}X^2Y^2+\ldots aX^3Y^3\ldots
\end{aligned}
$$

where X and Y are latitude and longitude coordinates for (approximate) geometric centres of the wards.

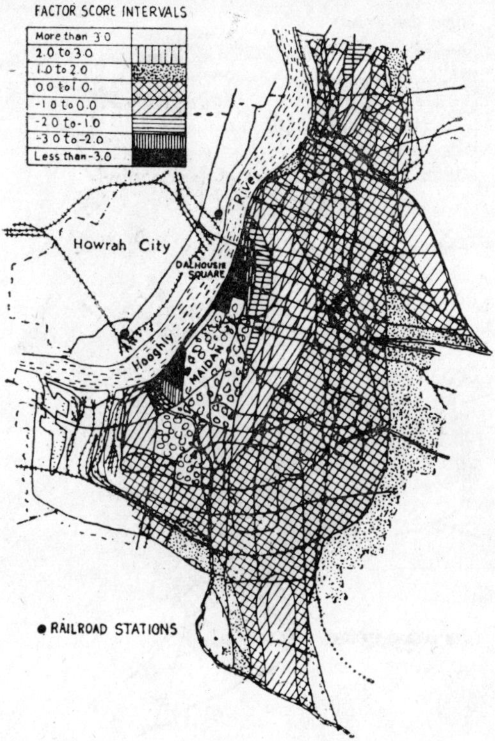

FIG. 4.3 : *Factor 1 : Land use and familism*

surfaces for some of the variables reveals how their commonalities are captured by this factor.

Factors 5 and 6 : traditional commercial communities and the peripheral ring

Surrounding the city core, two traditional ethnic commercial communities-cum-functional areas were identified by the factor analysis. Principal loadings on factor 6 were : percentage workers employed in trade, —0.560 ; percentage buildings in commercial use, —0.422 ; and percentage population in scheduled castes, 0.541. In the factor score map, areas of former Bengali commercial caste concentration in the southern half of the old native quarter stand out (Fig. 4.4) in stark contrast to the traditional peripheral ring of the lower-status scheduled castes.

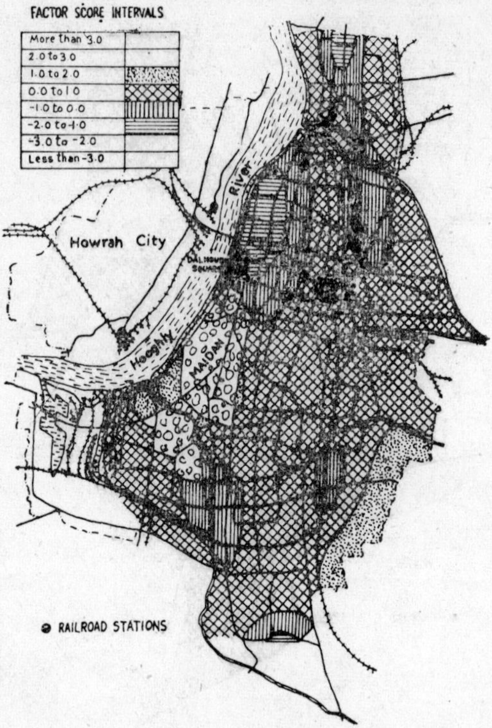

FIG. 4.4 : *Factor 6 : Traditional Bengali commercial community*

Variables loading on factor 5 were the following : percentage of buildings in mixed residential-commercial use, 0.825 ; percentage of buildings in business and offices, —0.298 ; percentage of population Hindu, —0.678 ; and percentage of population non-Bengali, 0.303. This cluster of variables indexes the relative concentration of the non-Bengali commercial classes in the wards immediately east of the Maidan.

The zones identified by factors 5 and 6 are not simply traditional ethnic community areas, however. They occupy the principal zones of commercial growth between 1911 and 1961, so that if their ethnic bases are weakening, as Bose suggests, their functional role in the city is becoming more marked, as areas of specialized land use.

Factor 4 : the substantial residential areas

The variable configuration and factor scores isolated by the fourth dimension identify the most substantial residential areas of Calcutta. Factor loadings are as follows : percentage of children under five, 0.778 ; percentage of houses with more than one room, −0.784 ; percentage of houses with walls and roof, 0.556 ; percentage of houses with mud walls, 0.361 ; percentage of population literate, —0.494 ; percentage of population in scheduled castes, 0.357 ; percentage of workers employed in manufacturing, 0.465 ; percentage of workers employed in services, —0.357 ; and percentage of buildings schools, —0.392.

Two zones are identified : the Bengali upper-caste northern half of the old native quarter and the Europe residential zones east of the Maidan, with the latter rating somewhat higher than the former. Over the years, as their ethnic base has weakened, those areas have increased their residential function.

Somewhat higher literacy rates than elsewhere and lower proportions of children under five (an inverse relationship noted in transitional communities elsewhere) are among the features that distinguish the residents of these two areas of larger, sounder homes from more peripheral residential wards in which there are greater proportions of mud-walled single roomed bustees housing higher proportions of manufacturing workers. Thus, the factor apparently comes close to indexing status.

Factor 3 : axiality in literacy

However, a substantial variation in percentage of literate males (loading 0.808) and females (0.917) overlap the city's mixed and dominantly residential areas, especially in the areas of houses with more than one room (0.399) where the proportion of buildings that are places of education is greatest (0.566). The high literacy wards stretch along a southward axis from Park Street through Ballygunge to Tollygunge and include sections of Alipore. Characteristically, wards with the greatest concentrations of scheduled castes have the lowest literacy levels (—0.345) ; these are the wards with the greatest admixture of Oriya and Bihari in-migrants. This separation of factor 3 from factor 4 indicates that the links between status and literacy remain far from absolute in Calcutta.

Factor 2 : higher-status muslim concentrations

Loadings on this factor are : percentage of workers employed in manufacturing, 0.696 ; percentage of workers employed in construction, 0.807 ; percentage of workers employed in trade, 0.722 ; percentage of workers employed in transportation, 0.776 ; percentage of workers employed in services, 0.851 ; percentage of females employed 0.879 ; and percentage of houses with more than one household, —0.633. Interpretation of this configuration of variables is not particularly transparent, until the map of factor scores is studied simultaneously. Broad areas of the city have uniform levels of positive scores, but certain extreme negative concentrations stand out, identifying communities having many houses containing more than one household, low percentages of females employed, and so on. These communities are the principal areas of upper- and middle-class Muslim concentration.

Factor 7 through 10 : special land-use configurations

The remaining four factors with eigenvalues exceeding unity produced in the analysis identified special features of Calcutta's land use pattern :

(a) Peripheral, essentially rural, areas with substantially higher sizes of households than the rest of the city (0.838).

(b) Hotel-restaurant-entertainment districts (percentage of buildings commercial and restaurants (0.661) ; percentage of places of entertainment (0.869).

(c) The medical centre (percentage of buildings medical institutions (0.576).

(d) The factory zones (percentage of buildings factories (0.691).

These round out the meaningful variations extractable from the original 80×37 data matrix, and they further indicate the substantial functional differentiation of land use achieved in Calcutta by 1961.

Calcutta's social areas

One valuable outcome of the analysis was an 80×10 matrix of

orthonormal factor scores. Relative similarity of each ward with every other could thus be measured as the distance separating the wards when plotted as points in the Euclidean space of the 10 dimensions (Gower 1966). An 80 × 80 distance matrix portrays these similarities. Application of a grouping algorithm to this matrix permits the wards to be allocated to relatively homogeneous subsets, consistent with the notion of social areas.

In the present case, a stepwise grouping algorithm that successively minimizes within group variance was utilized. The city was found to subdivide into a set of sub-regions (social areas), as portrayed in Figure 4.5. In brief, these comprise the city core, the superior residential areas in the northern part of the native quarter and east of the Maidan, the mixed commercial residential areas of the southern half of the native quarter and southeast of the city core, and the broad ring of peripheral wards. The

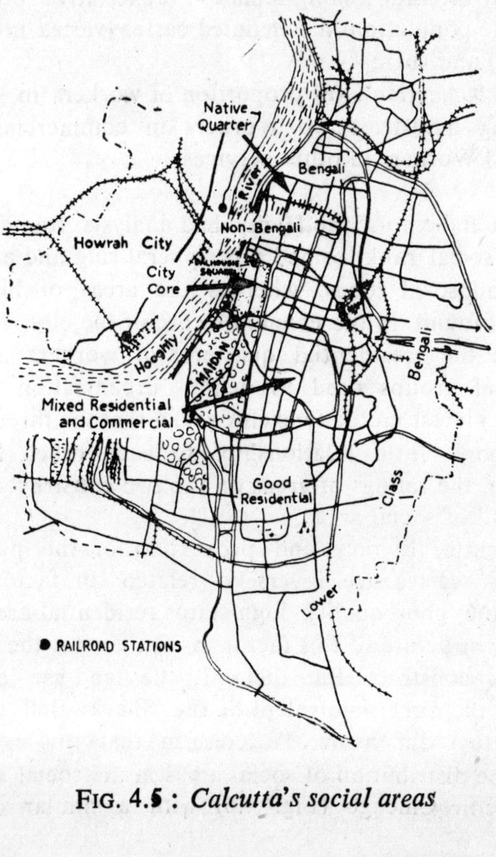

FIG. 4.5 : *Calcutta's social areas*

legacy of the past remained strongly imprinted in the social space of the transitional city of 1961.

DISCUSSION

The factorial ecology belies the assumption made in a social area analysis, stricto sensu of Hyderabad (1966); however, that the social space of an Indian city can be studied simply using Indian equivalents of the two principal Shevky-Bell constructs of social rank and urbanization. According to the Hyderabad research workers, the two constructs are constituted in the Indian case as follows :

1. *Social rank* : (a) Literacy among the general population ; (b) Literacy among females ; (c) Relative proportion of the population in scheduled castes (varies inversely with (a) and (b)) ;
2. *Urbanization* : (a) proportion of workers in manufacturing industries ; (b) Workers in commercial activities ; (c) Workers in other services.

Using these indicators, the Hyderabad analysts found that areas of lowest social rank formed a peripheral ring and also occupied the city's industrial zones, whereas the areas of highest social rank were found in the economic core of the city. But they also found that the residential quarters of workers in the three occupational groups used to define urbanization were highly segregated, almost mutually exclusive, so that the three constituent variables bore little relationship to each other. Nonetheless, they crossed the ratings of areas on the two constructs to develop four "types" of social areas.

In Calcutta, literacy and proportion of the population of "scheduled" castes are inversely related on factor 4, which separates the good-quality high-status residential areas from the others. The structure of this factor is similar to the Hyderabad social rank construct. But factor 1, the land use and familism gradient, is the direct equivalent of the Shevky-Bell urbanization (family status) dimension. To confirm that this is so we can compare the distribution of social areas in the social status-family status space in Chicago (Fig. 4.6) with a similar diagram for

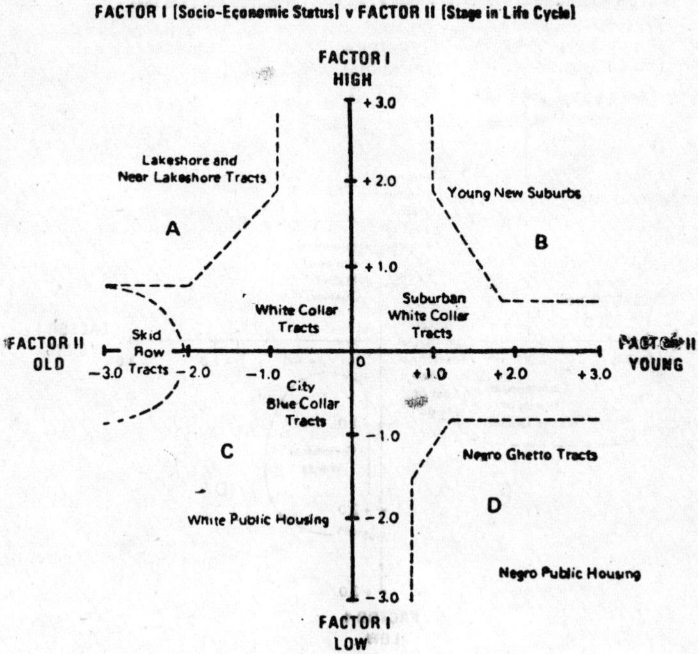

FIG. 4.6 : *Chicago's social space*

Calcutta in which factor 4 scores are used to locate wards on the ordinate and factor 1 scores to provide positioning with respect to the abscissa (Fig. 4.7). Similar segments of the social space are occupied by analogous social areas. In both instances, the high-status residential areas front superior amenities (lake in Chicago and Maidan in Calcutta). On the other hand, the geographic pattern of the other three social areas is inverted (Fig. 4.8).

Calcutta's social geography still thus contains many traditional elements. But, if, as Bose argues, regional or ethnic groups in Calcutta are coming to bear the same relationships to each other as do the castes in India as a whole, the ethnic bases of the communities are being supplemented by strong functional correlates. There is increasing functional differentiation of land use, alongside regional occupational differentiation. And consistent with the notion of a city in transition, alongside these bases of differentiation are a strong land use and familism gradient,

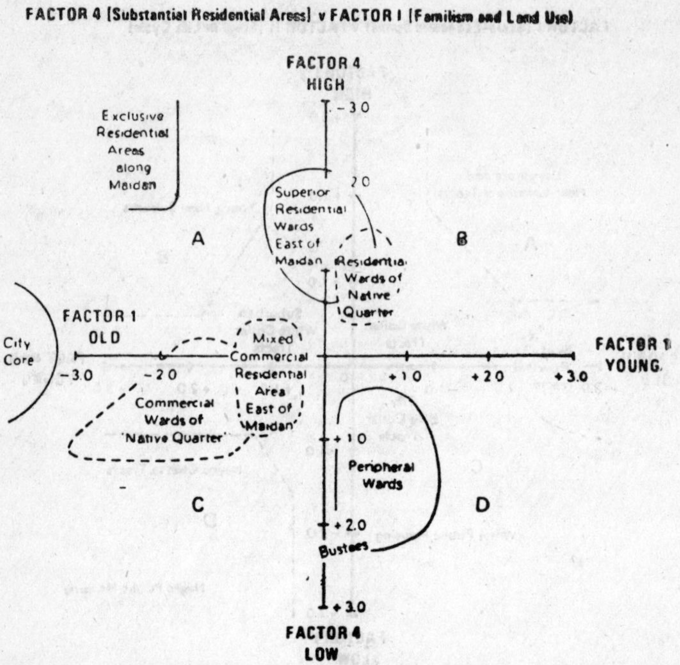

FIG. 4.7 : *Calcutta's social space*

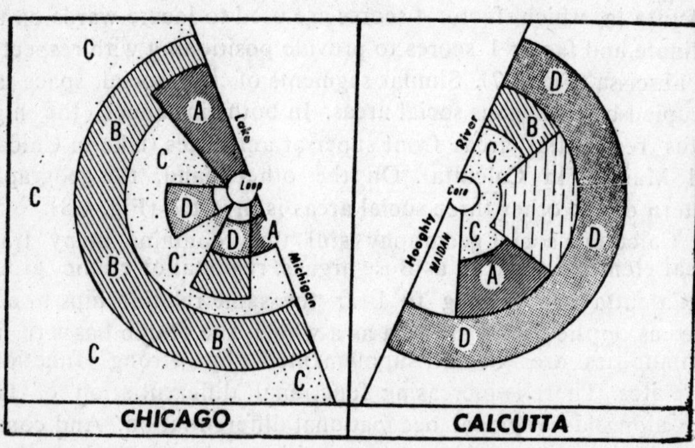

FIG. 4.8 : *Generalised social areas*

comparable in many ways to "modern" American formulations of urbanization. On the other hand, several dimensions are required to index status, because many of the ethnic linkages between occupation, region of origin, and caste remain—factor 6 differentiates the areas occupied by the Bengali "middle" commercial castes from those of the "scheduled" castes ; factor 5 is needed to define the middle-status non-Bengali commercial zones ; and factor 4 separates the high-status residential areas but does not completely differentiate the commercial wards from the periphery ; and factor 3 reveals the familiar "western" Hoyt-like axiality, but restricts it to literacy. Female employment is not part of the familism pattern, as in the Shevky-Bell model, but is related to the differences between Hindu and Muslim.

In terms of the factor models outlined in the fourth section of this paper, then, Calcutta conforms most closely to Combination Three, in which there is a separate family-status dimension, but socio-economic status and minority group membership are linked. It is of some interest that the closest resemblances of Calcutta's ecology are to the ecology of cities in the American south, where traditionally one found links between race and status in a system of caste. One description, written by Davis, Gardner, and Gardner (1941) reads :

> Life in the communities in the Deep South follows an orderly pattern. The inhabitants live in a social world clearly divided into two ranks, the white caste and Negro caste. The coloured castes share disproportionately in the privileges and obligations of labour, school, and government.

This particular factor combination raises serious operations about relative emphasis placed on different dimensions in the choice model outlined at the beginning of this paper. Bose noted the interrelationships between scarcity and ethnicity in Calcutta. The resulting ethnic base of much decision making can be embodied in the residential choice model with appropriate modifications, however. To the two choice dimensions of individual social space discussed in the American context can be added a third dimension of ethnicity (the regional relativity of the "American" model must by now also be evident). The dimensions can be re-ordered, placing ethnicity at the head and altered in length, making ethnicity the most important element in the choice process.

Thus, in Calcutta, a new Rajasthani or Punjabi in-migrant usually seeks the safety of his ethnic enclave. Because most non-Bengali ethnic groups also occupy particular occupational niches, the variance in status within ethnic groups is substantially less than that between ethnic groups. In effect, ethnicity overrides choice based upon status, for the status and ethnic dimensions are collinear, with the latter more fundamental in defining the social dimensions within which choices are made. The exceptions are within the Bengali community, where occupationally derived differences corresponding to traditional caste differentiation still persist.

Similar less-than-economic decisions are involved in the creation of bustees, emphasizing the parallel relativity of location of many of the ethnic enclaves in physical space. Around the periphery of the city are found the poorest residents of and migrants to the city: scheduled castes, refugees from East Bengal, and lower-caste non-Bengalis. The process at work in the creation of the bustees they occupy was not one of orderly choice in an ethnically governed market framework but rather one of occupation, often illegally, of "empty" land at the urban periphery by poor families. There they erect a shack and attempt to survive in incredibly over-crowded and unhealthy living conditions. Abrams (1964) has vividly described the sequence of events :

> As in a military campaign, some would bivouac during the night with their stock of materials behind a newly placed billboard. Next day, the horizon would be dotted with new rows of hovels, to which others would be added shack by shack, until the expansion was checked by a road, by a canal or by an owner prepared to spill blood.

With the passage of time, the bustee changes from a frontier community into a "regular", market-involved urban slum. Residences are bought and sold, and changing status is reflected in either physical improvement of the dwelling or a move to a new place of residence. Such changes are constrained, however, by the low status accorded the bustee dwellers by their ccupations.

As the transitional process continues, however, one might reasonably expect the socio-economic and ethnic bases of differentiation to separate. The expectation is, therefore, that differing urban ecologies related to differing factor combinations can be arranged along a scale of urban development from pre- to post-industrial forms, but testing of this notion awaits systematic comparative ecological analysis. On the other hand, this systematization may have been advanced by the formalization induced by the factorial approach.

FIVE

BRIAN J.L. BERRY AND HOWARD SPODEK

COMPARATIVE ECOLOGIES OF LARGE INDIAN CITIES*

FACTORIAL interpretations of the ecology of Indian cities cannot be cast into the narrow conceptual mould of studies of American cities, for they must contend with a differing form of social differentiation, complex intermixture of occupation and ethnic background, substantially less family-type specialization, differing prevailing models of urban technology, differing historical attitudes to the city, and contrasting perceptions of the amenities of sites within it. An initial attempt to come to terms with the complexities of these differences has been provided by Berry and Rees (1969). This paper seeks to extend the exploration, by examining a group of cities (Ahmedabad, Bombay, Kanpur, Madras, Poona, Sholapur) at a point in time (1961), one of these cities (Poona) over a considerable time-span (1822-1954), and one (Bombay) at a range of observational scales. A primary concern is to penetrate more deeply into traditional styles of Indian culture as they affect residence patterns.

The caveats are predictable. Data are very uneven. The materials on Poona and Sholapur were produced in social surveys under the direction of D.R. Gadgil at the Gokhale Institute of Politics and Economics in Poona and are quite good (Gadgil 1952, 1965). Data for the other cities are derived from the Census of India (1961 a, b, c, d), which has a somewhat higher

* Reprinted from *Economic Geography*, Vol. 47, No. 2 (supplement), 1971, by permission of the authors and editors.

degree of error. The Census operation was largely in the hands of state government bureaus responsible to the Central Government Census Office and these state governments collected different varieties of data. Thus the data for Ahmedabad are richer than those for Madras, which are in turn somewhat richer than the materials available for Bombay and Kanpur. In addition, the census tracts within and among cities differ widely in size, from a few hundred residents to tens of thousands, and very obviously, also in internal homogeneity.*

Data over time for all cities except Poona are unavailable since the Indian census only began reporting data for cities by wards in 1961. Also, the number of cities covered is severely limited. The 1961 census promised sections from each state census office treating the large cities of the state. For reasons of non-compliance with this directive, or more simply of delay, most of these city volumes have yet to appear. Some data on a ward basis are available in the census volumes on the districts of India, but even where such data were available, as for Poona in 1961, there were no accompanying maps to indicate which ward was which. Finally, much data on occupation is gathered on a vertical industrial basis, that is, the classification of workers by industry included everyone from owners of large factories to machine operators, to factory scavengers, to apprentices in three-man workshops. All these obstacles contribute to making this study still a preliminary one. But it was felt that it would be worthwhile to proceed, despite the difficulties, since these were all the systematic data on Indian cities likely to be available for some years to come.

TRADITIONAL INDIAN RESIDENCE PATTERNS

The empirical part of this study uses factorial methods, results of which have no ready interpretations without an adequate

* In the case of Ahmedabad, tests were undertaken, using analysis of variance, to assay the relative variability between and within the census tracts. Lacking variance estimates for individuals, internal variability was measured by block-to-block variations within wards, with an average of 20 blocks per ward. As a case in point, female literacy produced an F of 26.7 (28 and 454 d.f.); the greatest source of variability remained that between the observational units.

contextual base. Fortunately, there are at least two sources that give a picture of normative patterns of residence in India. The first is ancient religious and policy-making texts, the second is residence patterns in village India.

The ancient texts

The texts, dating from perhaps 300 A.D. back to 500 B.C. and covering both the northern Aryan and southern Dravidian areas of India, represent both texts on political management like the *Arthasastra* and texts on architecture and design such as the *Agni Purana*, the *Manasara*, and the *Sukranitisara.* These texts indicate the antiquity of urbanization of India, and the types of towns which characterized the subcontinent in ancient times. They included princely cities, temple centres, market places and trading posts, university towns, industrial centres, and nucleated towns formed by the unions of two or more villages. But the essential features of all these types of cities included a fortress, religious institutions, and socio-economic separation of the people into distinct neighbourhoods or "natural areas" (Ghurye 1961). In the Vedic literature the word *pura*, the modern Sanskrit synonym of town, was used in the sense of a fort or stronghold (Dutt 1925). The major towns were associated with capitals of local rulers. "Because India was not brought under the suzerainty of one single emperor, but on the contrary, was a medley of small principalities fighting with one another for overlordship, the military camps were turned into royal capitals. Indeed, each prince had to build a new city in order to demonstrate his grandeur and not to be ensnared in the many byways in the former capital which were better known to the former rulers than to himself" (Dutt 1925). In the Dravidian south, the Tamil word *nagar* carried at least five different meanings : house, temple, palace, castle, city (Ayyar 1916).

The physical structure of the ancient town reflected its fortress qualities in walls and moats, its royal and administrative buildings either in the centre of the town or in the northern quadrant, and temples of the city's tutelary deities, again either in the city centre or near the corners of the walls. Most striking was the division of people into distinctly separate quadrants and neighbourhoods on the basis of caste or occupational status. For example, the *Arthasastra* describes the kind of frontier town which the ruler should build ; the varna-castes are clearly divided into separate parts of town. Brahmins, the priests and teachers, are grouped in the northern section, around the royal palace ; Kshatriyas, the military, are in the east along with some traders ; the trading and landholding castes of Vaishyas are in the south ; and the Sudras or working classes are in the west. Outcastes are totally outside the city walls, near the cremation grounds.

The *Agni Purana* presents a very similar picture of the division of peoples, although there are significant differences. The temples are not at the centre of the town, but at the corners of the walls serving to protect the town spiritually. The shops do not flank the central shrines, but are located near, but not at, the four corners of the town. The main four caste-varna groupings are still divided in the four directions as above, though the agricultural traders are now to be grouped with Brahmins ; many professionals are grouped with the Sudras ; and dancers, musicians, and prostitutes have been joined with the trading castes in the south. This last combination is not so unusual since each of these occupations involves contact with diverse peoples and occupations and the provision of services which the dominant culture often regards with mixed emotions.

Ayyar's representation of the form of Madura in South India indicates that this town had similar features in ancient times, but the bulk of the population is left unrepresented. Thus the palace and court retinue are located in the northeast ; the religious and educational establishment has its own quarter, though it is now to the east ; and the marginal people such as musicians, artists, and dancers are grouped with the prostitutes at the other side of the city to the west. The bazaar area and its craftsmen and merchants are to the south just inside the main gate to the fortified city. No mention is made of the other inhabitants of the city, nor of the outcaste areas, though presumably the latter, as in other ancient cities are outside the gates.

Was such a rigid form of residential patterns ever employed ? Probably not. The projected plans actually indicate not what actual city planning as a socio-economic phenomenon was like in contemporary times, but more a kind of abstraction following the mechanical set-up of different occupational groups, castes, and classes in the city with gods, kings, and priests as the centre of the whole scheme. And the plans for different cities, or by different authors, indicate differential groupings, thereby revealing the extent to which values and perceptions can and did vary. The most pronounced characteristic, however, the separation of the court, the Brahmins and the outcastes into distinct areas, separate from each other and from other elements of the population, seems common throughout. The *Sukranitisara*, one of the architectural treatises thus specifically groups only the royal palace, court council buildings, museum, officials, and clerks at the centre of the city and the untouchables alongside the cremation grounds on the outskirts.

Caste : classical and modern

Clearly, a fundamental determinant of the locational prescriptions in the texts is caste. The classical textual division of people was that of *varna-ashrama dharma* ; the India-wide separation of people into four groupings : Brahmin or teacher-priest, Kshatriya or warrior-prince, Vaishya or wealthy landholder and trader, Sudras or workers, plus outcastes. It is quite nuclear that such a four-fold division of society ever existed in social fact ; indeed, in modern functional terms, caste is very different from the system presented in the classical texts. The modern view of the Indian caste system which has been developed largely by anthropological work in India over the past twenty years, sees caste groups—*jatis*—as limited in territory and membership so that far from there being only four all-India castes, there are instead thousands of castes, of which anywhere from two or three to forty or fifty may be represented in any one area. Membership in a caste comes with birth, but does not necessarily, determine occupation. It will most likely determine the limit of eligible people for marriage, and it will significantly affect the scope of social and commensal activities. Caste is hierarchical, but not rigidly so. Thus in any one area it is likely that the top and bottom castes will be distinguishable with relatively little difficulty, but the rankings of intermediate castes may be debatable.

Caste and class are antithetical in that the latter system is said to have a rather high degree of potential socio-economic mobility while the former has relatively little. Caste also is more a village phenomenon than an urban one since in village societies the essential base of wealth and power is in only one sphere, possession of land, while in cities there are many areas in which a man can achieve wealth and power. Policy morphous urban society shakes loose the rigidities of the caste system. One of the most significant variables which the urban ecologist studies everywhere in the world, the social basis for residential patterns, therefore, takes on special interest in India for it relates closely to the power of the caste system in determining residential patterns.

Residence patterns in village India

Since India is an urbanizing society in which the early socialization of many urban dwellers still takes place in villages, the role of caste in residence patterns in these village settings is of some importance. Mayer's perceptive study of one village (Mayer 1960) is a useful starting place :

> The village does not have streets whose entry is restricted, even in the wards occupied by the Harijans (outcastes). In general there is a tendency for castes of roughly equal status to inhabit the same locality of the village. All over there is a certain amount of intermingling though it is clear which caste provides the nucleus in any ward. But in the Harijan wards there is clear separation both from other castes and between the Harijan castes themselves. The sweepers are to one side of the village and the tanners to the other. The weavers form a more dispersed pattern ranging behind the houses of the Rajputs who were, and still to some extent are, their masters.

From further south, in the State of Mysore, Epstein (1962) presents maps of two villages. In both villages, the groups who are considered outsiders are clearly located on the fringe in their own groupings : untouchables, Muslims, washermen, and even public works department personnel not native to the village. The other castes and lineages are somewhat more mingled, but the leading lineages including that of the headman tend to have a central location on the most solid and respectable street. Still, among the Hindu castes there is somewhat more mingling in the residence patterns, so that while richer peasants dominate the central area, the artisan groups for the most part live mingled among the other peasant lineages.

Another village of Mysore State, Aminbhavi, is described by Spate and Learmonth (1967) :

> caste and community largely govern the layout . . . Each caste tends to occupy a solid block of contiguous houses in a lane named from the caste. Low castes and outcastes live on the fringes of the village and even beyond the old moat.

The authors add a note about the high correlation between caste status and occupational status. "Occupations likewise are still mainly on a caste basis : the Lingayats provide the bulk of the tenant-farmers, Talwars and Harijans landless agricultural labour ; carpenters, smiths, cobblers, washermen, barbers are all separate castes."

From these examples, a few general patterns are observable. The centre of the village is usually pre-empted by the wealthiest and most prestigious caste groupings, as denoted by the lineage of the headman. Outcastes, untouchables, non-Hindus, and non-natives of the village are located on the periphery of the village, or even outside of it. Other castes often form separate neighbourhoods of greater or lesser exclusivity and physical separation from one another, but castes which have only one or two representative families usually locate among more numerous castes of approximately equal ritual, social, and economic status. Both caste membership and class status are thus ingredients in residence patterns, particularly at the extreme ends of the hierarchy.

Residence patterns in a small town

An intermediate case between the large cities analyzed below and the village studies is provided by Fox's monograph on Tezibazar, a town of some 7,000 inhabitants in eastern U.P. He found that "no necessary caste or communal patterning to residential areas exists except for untouchables. Muslim and Hindu, Brahmin and Baniya castes all live intermixed or interspersed" (Fox 1969). In the past, however, neighbourhoods may have been built along a caste basis. What has caused the change ? Fox argues that it is the development of business and a political change which has led to the passing of control from the hands of the hereditary

Zamindar or landed proprietor to freer and more open competition for economic power in the commercial sphere. Now, especially in the newer residential quarters, economic class characteristics predominate. "Even in those relatively non-commercial and peripheral sections of the town where some caste-neighbourhood congruence exists, it is a residue of the past rather than a presently significant social patterning of town society. This situation is true for all castes and communities other than untouchables" (Fox 1969). In this town, then, the expected passing of caste-based neighbourhoods to those based on economic class is taking place as commerce becomes more lively. Families of various castes come together to live in neighbourhoods of a common economic status. Again, however, as in village India, the untouchables remain unintegrated at least in terms of residue into this new class structure.

ECOLOGY OF THE LARGE INDIAN CITIES

Armed with the results of the earlier Berry-Rees (1969) study, which showed interpenetration of classical and modern elements in the social geography of Calcutta, and a broader contextual base, this analysis of six major cities was designed to explore further the intricacies of India's urban ecology.

Five major categories were chosen as spanning the variables most likely to provide answers relating to these questions concerning the social and physical patterns of India's urban residents. Data were gathered from available sources on : density-distance from centre of town ; demographic variables ; caste, religion, and ethnicity (i.e., ascriptive status) ; socio-economic status including education, occupation, and income ; and housing data. All cities had at least some data in each category, though the types of data and the amounts varied from city to city.

Since, as in the Berry-Rees study, the primary purpose of the data analysis was an exploratory one of pattern recognition, no elaborate examination of alternative factoring procedures was undertaken. Instead, all the results are based upon principal axis factor analyses with normal varimax rotation and pseudo-inverse orthonormal factor scores. Data were transformed where necessary to satisfy linearity assumptions. A cascading sequence of rotations was undertaken in each case from initial principal axes, beginning with the first pair and then by steps for the first three axes, the first four, and so on, until the final rotated solution included all principal axes with eigenvalues exceeding unity.

Tables 5.1 to 5.4 summarize the patterns common to several of the cities and to the earlier Calcutta study : a socio-economic status factor, a particular index of family-type differentiation, and a communal factor or factors. To the extent that these are repetitive, ingredients in the residential systems of the cities under consideration, they may be thought to be general properties of India's urban ecology. Tables 5.5 through 5.14 present the factor matrices for the final rotated solutions for each case.

TABLE 5.1 : **General factor patterns**

Factor	General pattern (Sign of factor loadings)	
Socio-economic status (SES)	Male literacy	+
	Female literacy	+
	Males in Trade (%)	+
	Scheduled caste (%)	—
	Males in Manufacturing	—
Familism	Sex ratio	+
	Working males	—
Communal	Brahmin (%)	+
	Muslim (%)	—
	Scheduled caste (%)	—
	Maratha (%)	+
	Non-Maratha (%)	—
	Lingayat (%)	—

Status differentiation

The cluster of variables defining high socio-economic status (SES) in the analyses comprised high rates of male and female literacy, a high percentage of males engaged in occupations of trade and commerce, and, conversely, low percentages of scheduled caste members, and of males in manufacturing. Of these variables, nine cities had available data (Poona in 1822 did not), and eight of them showed this factor as significant in their residential

TABLE 5.2 : **Cities with a socio-economic status factor**

City	Persence as factor	Additional variables	
Poona, 1822	See discussion of communal factor		
Poona, 1937	Factor I	Density factors	+
		Brahmins (%)	+
		Working females (%)	—
		Non-working (%)	+
		White-collar (%)	+
		Industry	+
		Shops	+
Poona, 1954	Factor I (no data on percentage males in manufacturing)	Working males (%)	—
		Working females (%)	—
		Income per capita	+
		Muslim (%)	—
		Non-Maratha (%)	—
		Brahmin (%)	+
		Houseless (%)	+
Bombay (15 observations)	Factor III		
Bombay (88 observations)	Factor I	Density	+
Bombay (437 observations)	Factor I	Working females (%)	—
Kanpur	Factor I		
Sholapur, 1938	See discussion of communal factor		
Ahmedabad	Factor I (no data on percentage of males in trade ; manufacturing does not correlate + or —)	Density	+
		Females/males	+
		Working males (%)	—
		Working females (%)	—
		Females in household industry (%)	+
		Huts (%)	—
Madras	Factor I (except percentage of males in manufacturing)		

patterns.

The additional variables which correlate with high socio-economic status in the cities where the variables are available, for the most part, are quite expectable : high proportions of white collar workers, high proportions of non-workers and of women who do not work, high incomes per capita, and low

TABLE 5.3 : **Cities with a familism index**

City	Presence as factor	Additional variables	
Poona, 1822	No data		
Poona, 1937	Family survey (data not applicable)		
Poona, 1954	Family survey (data not applicable)		
Bombay (15 observations)	Factor I	Institutional population	—
		Houseless (%)	—
		Literate males (%)	—
		Literate females (%)	+
		Working females (%)	—
		Males in manufacturing (%)	+
		Males in transportation (%)	—
Bombay (88 observations)	Factor III	Literate females (%)	+
		Houseless (%)	—
		Density	—
Bombay (437 observations)	Factor II	Males in manufacturing (%)	—
		Literate females (%)	+
Kanpur	Factor II	Literate males (%)	+
Sholapur	Family survey (data not applicable)		
Ahmedabad	Factor VI	Males in manufacturing (%)	—
		Residential (%)	—
		Factory (%)	—
		Business offices (%)	+
		Shops	+
	Factor I	Correlates with S.E.S.	
	Factor VII	Males in household (%)	+
		Males in construction (%)	+
		Females in household (%)	+
		Factory (%)	+
		Shops (%)	+
Madras	Factor III	Houseless (%)	—
		Males in manufacturing (%)	+
		Males in trade (%)	—

proportions of huts.

Other parts of the output point to the strong link between caste status and economic status. Wards which had high SES

TABLE 5.4 : **Cities with a communal factor or factors**

City	Presence as factor	Additional variables	
Poona, 1822	Factor I	Brahmin (%)	+
		Muslim (%)	—
		Persons per house (%)	+
		1-storey houses (%)	—
		Huts (%)	—
	Factor II	Brahmin (%)	+
		Scheduled caste (%)	—
		Total population	+
		Persons per house	+
Poona, 1937	Factor I (correlates with S.E.S.)	Brahmin (%)	+
		Scheduled caste (%)	—
	Factor IV	Brahmin (%)	+
		Muslim (%)	—
		Persons per house	—
		Non-Maratha (%)	—
		Non-working (%)	+
		Owner resident (%)	+
		Area	+
Poona, 1954	Factor I (correlates with S.E.S.)	Brahmin (%)	+
		Muslim (%)	—
		Scheduled caste (%)	—
		Non-Maratha (%)	—
Bombay (15 observations)	No data		
Bombay (88 observations)	No data		
Bombay (437 observations)	No data		
Kanupr	No data		
Sholapur	Factor I	Brahmin (%)	+
		Maratha (%)	+
		Rooms per family	+
		Square feet per person	+
		1-storey houses (%)	—
		Families below poverty line (%)	—
		Families below destitution line (%)	—
		Income per capita	+
		Literate males (%)	+
		Literate females (%)	+
		Working males (%)	—
		Doctors and lawyers (%)	+

TABLE 5.4 : (*Contd.*)

		White-collar (%)	+
		Groceries	+
		Hardware	+
		Single males (%)	+
		Single females (%)	+
	Factor II	Lingayat (%)	—
		Non-Maratha (%)	—
		1-storey houses (%)	+
		Square feet per person	+
		Literate males (%)	+
		Working females (%)	+
		White-collar (%)	—
		Unskilled (%)	—
		Families below poverty Line (%)	+
		Incomes per capita	—
		Groceries	—
		Hardware	—
	Factor III (weak correlation)		
Ahmedabad	No data		
Madras	No data		

were the same wards which had high proportions of high caste people, Brahmins, and relatively low proportions of Muslims and migrants (non-Marathi in Poona). This may or may not indicate an identity between high caste status and high economic status, but it does indicate that many wards which are high on one index are high on the other. Other loadings link SES with land use patterns. High status areas are located in the centre of town rather than in the suburbs, substantiating the ideas of Sjoberg (1960).

Maps of the factor scores (not reproduced here because of space limitations) show spatial distributions of SES readily explainable by the historical and economic development of the cities. Poona, for example, developed early around a fort centre on the Mutha river. To the east and south of the fort was the commercial centre of town, and this area became the chief location for Muslims, non-Marathis, and the commercial life of Poona. To the west of the fort, filling the area between the Mutha river and a vanished river which followed the course of today's north-south Katraj aqueduct, lived the Brahmins of the town. Poona has had for centuries a most distinguished and large

TABLE 5.5 : **Factor structure—Ahmedabad (34 observations)**

Variable	Communality	Factor I	II	III	IV	V	VI	VII
Density	0.818	—0.669	0.482	—	—	—	—	—
Persons per family	0.782	—0.317	0.736	—	—	—	—	—
Total population	0.620	—	—	—	0.631	—	—	—
Females/males	0.829	—0.343	—	0.430	—	—	—0.561	—0.430
Scheduled caste (%)	0.828	0.854	—	—	—	—	—	—
Literate males (%)	0.927	—0.703	—	0.452	—0.438	—	—	—
Literate females (%)	0.887	—0.733	—	0.357	—0.432	—	—	—
Working males (%)	0.808	0.389	—	—	—	—	0.644	0.304
Working females (%)	0.888	0.728		—0.501	—	—	—	—
Males in farming (%)	0.938	—	—	—0.948	—	—	—	—
Males in transportation (%)	0.777	—	—	—0.830	—	—	—	—
Males in manufacturing (%)	0.774	—	—	—	0.735	—	—0.361	—
Males in household (%)	0.761	—	0.319	—	—	—	—	—0.773
Males in construction (%)	0.760	—	—	—	—	—	—	—0.768
Females in farming (%)	0.893	—	—0.363	—0.820	—	—	—	—
Females in transportation (%)	0.961	—	—	—	—	0.950	—	—
Females in manufacturing (%)	0.767	—	—	—	0.801	—	—	—
Females in household (%)	0.836	—0.545	0.519	—	0.350	—	—	—0.360
Females in construction (%)	0.969	—	—	—	—	0.920	—	—
Huts (%)	0.782	0.859	—	—	—	—	—	—
Vacant (%)	0.833	—	—0.795	—	—	—	—	—
Residential (%)	0.780	—	—	—	0.350	—	—0.772	—
Restaurants (%)	0.741	0.347	—0.399	—	0.643	—	—	—
Factories (%)	0.756	—	0.404	—	—	—	—0.651	0.327

Business offices (%)	0.644	—	0.419	—	—0.316	—	0.596	—
Shops (%)	0.469	—	—	—	—	—	0.470	0.336
Males in trade (%)	0.807	—0.623	0.436	—	—	0.900	—	—
Distance	0.731	—	—0.750	—	—	—	—	—
EIGENVALUE	29Factors	17.4	29.7	41.3	52.4	62.4	72.1	80.3
	7Factors	21.6	36.9	51.4	65.3	77.7	89.7	100.0

TABLE 5.6 : **Factor structure—Bombay (15 observations)**

Variable	Communality	Factor I	II	III	IV
Total households	0.971	—	0.978	—	—
Density	0.935	—	0.366	—	—0.838
Total population	0.980	—	0.984	—	—
Institutional population	0.867	0.687	0.603	—	—
Males/females	0.960	—0.855	—0.399	—	—
Scheduled caste (%)	0.430	—	—	0.400	0.451
Houseless (%)	0.942	0.871	—	—0.401	—
Literate males (%)	0.953	0.381	—	—0.818	—
Literate females (%)	0.990	—0.316	—	—0.937	—
Working males (%)	0.896	0.910	—	—	—
Working females (%)	0.912	0.505	—	—	0.772
Males in manufacturing (%)	0.930	—0.379	0.375	0.782	—
Males in trade (%)	0.914	—	—	—0.496	—0.796
Males in transportation (%)	0.676	0.786	—	—	—
EIGENVALUE	14 Factors	29.8	51.1	71.4	88.3
	4 Factors	33.8	57.9	81.0	100.0

TABLE 5.7 : Factor structure—Bombay (88 observations)

Variable	Communality	Factor I	Factor II	Factor III	Factor IV	Factor V
Total households	0.935	—	0.956	—	—	—
Density	0.583	—0.544	—	—0.415	—	—0.310
Total population	0.945	—	0.953	—	—	—
Institutional population	0.768	—0.327	0.571	—	0.402	0.324
Males/females	0.892	—	—	0.878	—	—
Scheduled caste (%)	0.580	0.678	—	—	0.311	—
Houseless (%)	0.814	—	—	—0.566	0.631	—
Literate males (%)	0.784	—0.797	—	—	—	—
Literate females (%)	0.856	—0.770	—	0.479	—	—
Working males (%)	0.872	—	—	—0.922	—	—
Working females (%)	0.819	—	—	—	—	0.851
Males in manufacturing (%)	0.775	0.478	0.568	—	—	0.471
Males in trade (%)	0.801	—0.737	—	—	—	—0.442
Males in transportation (%)	0.800	—	—	—	0.882	—
EIGENVALUE	14 Factors	21.4	41.0	58.7	69.9	80.2
	5 Factors	26.7	51.1	73.3	87.1	100.0

TABLE 5.8 : **Factor structure—Bombay (437 observations)**

Variable	Communality	Factor I	II	III	IV	V
Total households	0.876	—	—	—0.925	—	—
Total population	0.852	—	—	—0.907	—	—
Institutional population	0.763	—	—	—	—	0.844
Males/females	0.895	—	—0.908	—	—	—
Scheduled caste (%)	0.551	—0.721	—	—	—	—
Houseless (%)	0.763	—	—	—	—0.818	—
Literate males (%)	0.797	0.828	—	—	—	—
Literate females (%)	0.846	0.684	—0.551	—	—	—
Working males (%)	0.873	—	0.919	—	—	—
Working females (%)	0.657	—0.564	—	0.304	—	0.497
Males in manufacturing (%)	0.671	—0.315	0.369	—0.445	0.369	—0.317
Males in trade (%)	0.637	0.751	—	—	—	—
Males in transportation (%)	0.854	—	—	—	—0.901	—
EIGENVALUE	14 Factors	19.3	35.7	50.5	62.8	71.7
	5 Factors	27.0	49.8	70.5	87.7	100.0

TABLE 5.9 : **Factor structure—Kanpur (136 observations)**

Variable	Communality	Factor I	Factor II	Factor III
Density	0.671	—	—	—0.770
Females/males	0.782	—	—0.833	—
Scheduled caste (%)	0.617	—0.645	—	0.448
Literate males (%)	0.795	0.816	—0.356	—
Literate females (%)	0.847	0.890	—	—
Working males (%)	0.896	—	0.895	—
Working females (%)	0.538	—	—	0.692
Males in manufacturing (%)	0.702	—0.803	—	—
Males in trade (%)	0.845	0.690	—	—0.598
EIGENVALUE	9 Factors	35.0	54.8	74.4
	3 Factors	47.0	73.6	100.0

TABLE 5.10 : **Factor structure—Madras (100 observations)**

Variable	Communality	Factor I	II	III	IV	V
Persons in family	0.834	—	—0.642	—	—	0.570
Households in house	0.854	—	—0.606	—	—	0.645
Total population	0.697	—	0.825	—	—	—
Females/males	0.694	—	—	0.807	—	—
Scheduled caste (%)	0.704	0.773	—	—	—	—
Houseless (%)	0.744	—	—	—0.840	—	—
Literate males (%)	0.847	—0.852	—	—	—0.341	—
Literate females (%)	0.887	—0.849	—	—	—0.358	—
Working males (%)	0.692	—	—	—0.713	0.321	—
Working females (%)	0.712	—	—	—	—	0.786
Males in manufacturing (%)	0.718	—	—	0.457	—	0.595
Males in trade (%)	0.813	—0.530	—0.512	—0.507	—	—
Males in household (%)	0.863	—	—	—	0.915	—
Females in household (%)	0.865	—	—	—	0.909	—
Area	0.679	—	0.816	—	—	—
EIGENVALUE	15 Factors	17.6	34.1	49.9	64.9	77.4
	5 Factors	22.8	44.1	64.5	83.9	100.0

TABLE 5.11 : **Factor structure—Poona, 1822 (22 observations)**

Variable	Communality	Factor I	Factor II	Factor III	Factor IV
Persons per house	0.687	0.636	—0.515	—	—
1-storey houses (%)	0.538	—0.678	—	—	—
Total population	0.761	—	—0.715	0.497	—
Females/males	0.834	—	—	0.902	—
Scheduled caste (%)	0.558	—	0.688	—	—
Brahmin (%)	0.545	0.607	—0.369	—	—
Muslim (%)	0.670	—0.799	—	—	—
Huts (%)	0.724	—0.354	0.683	—	0.306
Distance	0.878	—	—	—	0.937
EIGENVALUE	9 Factors	22.3	43.5	56.7	68.8
	4 Factors	32.3	63.2	82.3	100.0

TABLE 5.12 : **Factor structure—Poona, 1937 (22 observations)**

Variable	Communality	Factor I	II	III	IV	V	VI	VII	VIII	IX
Density	0.790	0.382	—	—	0.621	—	—	0.351	—	—
Persons per house	0.818	—	—	—	—	0.832	—	—	—	—
1-storey houses (%)	0.684	−0.752	—	—	—	—	—	—	—	—
Total population	0.855	—	—	0.840	—	—	—	—	—	—
Females/males	0.874	—	—	—	—	—	—	0.897	—	—
Scheduled caste (%)	0.758	−0.682	—	—	—	—	—	—	0.429	—
Brahmin (%)	0.808	0.612	0.373	−0.469	—	—	—	—	—	—
Muslim (%)	0.767	—	−0.358	—	0.647	—	—	−0.323	—	—
Maratha (%)	0.786	—	—	—	—	—	−0.686	0.306	—	—
Non-Maratha (%)	0.715	—	—	—	0.750	—	—	—	—	—
Square feet per capita	0.891	—	0.893	—	—	—	—	—	—	—
Huts (%)	0.641	−0.343	—	—	—	0.503	—	—	—	—
Literate females (%)	0.874	0.349	0.701	—	—	—	—	—	—	−0.305
Literate males (%)	0.857	0.562	0.697	—	—	—	—	—	—	—
Working males (%)	0.855	—	−0.437	—	—	—	−0.570	—	—	—
Working females (%)	0.751	−0·429	−0.512	—	—	0.340	—	−0.312	—	—
Non-working (%)	0.748	0.668	0.344	—	−0.335	—	—	—	—	—
Professional and owners	0.692	—	0.668	—	—	—	—	—	0.400	—
White-collar (%)	0.681	0.771	—	—	—	—	—	—	—	—
Unskilled (%)	0.763	−0.829	—	—	—	—	—	—	—	—
Artisan (%)	0.834	—	—	—	—	—	—	—	−0.863	—
Numbers of doctors and lawyers	0.879	0.424	0.364	0.675	—	—	—	—	—	—
Owner residences (%)	0.787	0.337	—	—	−0.374	−0.638	—	—	—	—

Income per capita	0.861	—	0.879	—	—	—	—	—	—	—
Industry	0.862	0.356	—	0.810	—	—	—	—	—	—
Shops	0.789	0.382	—	0.695	—	—	—	—	—	—
Males 15+ (%)	0.877	—	—	—	—	—	—0.819	—	—	—
Distance	0.753	—	—	—	—	—	—	—	—	0.790
Area	0.808	—	—	0.703	—0.457	—	—	—	—	—
Persons per family	0.902	—	—	—	—	—0.613	0.436	—	—	—0.492
EIGENVALUE	30 Factors	16.5	30.9	41.7	50.3	57.7	64.9	70.2	75.0	79.9
	9 Factors	20.7	38.7	52.2	63.0	72.3	81.3	87.8	94.0	100.0

TABLE 5.13 : **Factor structure—Poona, 1954 (22 observations)**

Variable	Communality	Factor I	II	III	IV	V	VI
Density	0.678	—	—	0.772	—	—	—
Total population	0.632	0.419	—	—	—	0.588	—
Females/males	0.914	—	—	—	—	—	0.906
Scheduled caste (%)	0.855	—0.655	—	—	0.558	—	—
Brahmin (%)	0.940	0.923	—	—	—	—	—
Maratha (%)	0.680	—	—	—	—	—0.806	—
Non-Maratha (%)	0.805	—0.317	—	0.611	—0.361	0.331	—
Artisan (%)	0.827	—	—	0.310	—0.837	—	—
Muslim (%)	0.820	—0.830	—	—	—	—	—
Literate males (%)	0.871	0.799	—	—	—0.369	—	—
Literate females (%)	0.870	0.899	—	—	—	—	—
Working males (%)	0.927	—0.493	—0.790	—	—	—	—
Working females (%)	0.795	—0.520	—	—0.563	—	—	0.377
Owner residences (%)	0.819	—	0.360	—	—0.470	—	0.609
Income per capita	0.702	0.778	—	—	—	—	—
Males 15+ (%)	0.857	—	—0.842	—	—	—	—
Distance	0.801	—	—	—	0.636	0.548	—
Areas	0.748	—	—	—0.669	—	0.425	—
Persons in family	0.780	—	0.683	0.441	—	—	—
EIGENVALUE	19 Factors	27.4	39.1	50.7	62.2	72.0	80.6
	6 Factors	34.0	48.5	62.9	77.2	89.3	100.0

TABLE 5.14 : **Factor structure—Sholapur (27 observations)**

Variable	Communality	Factor I	II	III	IV	V	VI	VII	VIII	IX
1-storey houses (%)	0.830	−0.371	0.607	−0.303	—	—	—	0.373	—	—
Females/males	0.870	—	—	—	−0.350	0.795	—	—	—	—
Scheduled caste (%)	0.827	—	—	−0.396	—	—	—	—	0.740	—
Brahmin (%)	0.934	0.897	—	—	—	—	—	—	—	—
Lingayat (%)	0.904	—	−0.913	—	—	—	—	—	—	—
Christian (%)	0.864	—	—	—	0.875	—	—	—	—	—
Muslim (%)	0.714	—	—	—	—	—	—	—	—	0.734
Maratha (%)	0.916	0.335	—	—	—	—	−0.863	—	—	—
Non-Maratha (%)	0.793	—	−0.597	−0.324	—	0.335	—	—	—	—
Rooms per family	0.827	0.717	—	—	—	—	—	—	—	—
Square feet per family	0.970	—	—	0.896	—	—	—	—	—	—
Square feet per person	0.948	0.364	−0.421	0.711	—	—	—	—	—	—
Huts (%)	—	—	—	—	—	—	—	—	0.859	—
Literate males (%)	0.823	0.558	−0.362	0.315	—	—	—	—	—	—
Literate females (%)	0.940	0.888	—	—	—	—	—	—	—	—
Working males (%)	0.860	−0.559	—	−0.319	—	—	—	−0.608	—	—
Working females (%)	0.776	—	0.344	−0.338	—	—	0.324	—	—	—
Number of doctors and lawyers	0.901	0.880	—	—	—	—	—	—	—	—
White-collar (%)	0.894	0.467	−0.524	0.467	—	—	—	—	—	—
Professionals and owners (%)	0.587	—	—	—	—	—	−0.606	—	—	—
Unskilled (%)	0.795	—	0.317	−0.425	0.391	—	—	—	0.497	—
Artisan (%)	0.843	—	—	—	—	—	—	—	—	0.901

TABLE 5.14 : (*Contd.*)

Variable	Communality	Factor I	II	III	IV	V	VI	VII	VIII	IX
Families below poverty line (%)	0.921	—0.814	0.381	—	—	—	—	—	—	—
Families below destitution line (%)	0.893	—0.643	—	—	—	—0.303	0.481	—0.359	—	—
Income per capita	0.858	0.514	—0.300	—	0.306	—	—0.453	—	—	—0.334
Groceries	0.890	0.633	—0.574	—	—	—	—	—	—	—
Hardware	0.906	0.343	—0.858	—	—	—	—	—	—	—
Industry	0.896	—	—	—	—0.357	—0.628	—	0.389	—	—
Males 15-45 (%)	0.767	—	—	—	—	—	—	0.821	—	—
Females 15-45 (%)	0.827	—	—	—0.510	—	—0.664	—	—	—	—
Single males (%)	0.900	0.805	—	—	—	—	—	—	—	—
Single females (%)	0.909	0.547	—	—	0.748	—	—	—	—	—
Persons per family	0.900	—	—	0.597	—	0.364	0.339	—0.457	—	—
EIGENVALUE	33 Factors	22.3	34.6	44.4	51.8	59.3	66.4	73.3	79.7	86.0
	9 Factors	25.9	40.2	51.6	60.2	68.9	77.2	85.2	92.7	100.0

group of Brahmins, running to twenty-five per cent of the city's population. During the days of the Maratha Empire, the rulers of the region instituted policies of great liberality towards Brahmins both as religious leaders and as members of government. Even after the fall of the Maratha Empire to the British, the city continued as a centre of Brahmin strength and a major educational centre with a number of all-India institutions. The map of Poona in 1822 shows the east-west split largely in communal terms, but by 1937 the split between the two sides of town was definitely one of socio-economic status.

The pattern of Madras is similar, though it was a city built from a mere cluster of villages by the British into a major metropolis of some two million inhabitants. The area around Fort St. George, the centre of British trade and administration from the seventeenth century onward, has become the hub around which are found the high SES wards. Another centre built centuries ago is Santhome and Mylapore, the areas where the Portuguese founded their fort in the late sixteenth century, and it, to this day, remains a high SES area. Finally, a high status zone in the "new residential" area indicates the beginning of suburbanization in Madras city (Dupuis 1960). In Madras, the areas of low SES correspond largely to the factory areas in the north, which abut the railroad track, and in the harbour area. The correlation of factory area and low SES residences is obvious and indeed forms a later factor in the analysis.

Ahmedabad, Kanpur, and Bombay also show higher socio-economic status toward the centre of the city, with lower status areas farther out, usually associated with the presence of large industrial areas. In Ahmedabad, the high status areas are all, except one, located in the old city which was surrounded by a wall until the 1920s. The low status areas were, it seems, always outside the old city and indeed, the wealthier citizens wanted it that way. "In a list of municipal wants put forward in the annual report of the Municipal Commission for 1887 to 1888 there was included : The removal of low caste and other such people for the reduction of overcrowdedness" (Gillion 1969). The banishment of low caste people from the downtown was a norm for urban India well before large-scale industrialization. Yet suburbanization has also come to Ahmedabad. The statistics in the fourteenth ward, the large sprawling ward across the

Sabarmati river from the old city, do not reflect the coming of increasing numbers of upper middle-class, white collar workers, or the establishment of the Gujarat University and of the state governmental secretariat in this area ; this is because the ward is so large that it includes diverse elements of population in addition to the upper middle-class suburbanites, averaging them out in the census reports.

Kanpur, similarly, has its high SES areas toward the centre of the old city. In addition, the residential section of the cantonment where the British established their homes apart from the Indians, and which has been taken over since Independence as a suburban development of the wealthy Indians, is a second, high status community. Some outlying areas are also being developed as a suburban, upper middle-class settlement. The low status neighbourhoods tend, again, to be located around the great industrial complexes which are the lifeblood of Kanpur's economy.

Bombay, too, has its areas of highest SES near the centre of town, adjacent to the commercial hub of the city. Conversely, low SES areas are near the ports, the industrial complex, and in outlying areas. For Bombay, data on increasingly fine-grained geographic units of observation were available. The largest units, *15-wards*, have populations from 58,000 to 660,000, and are as large as whole cities. It was with these units that the first generalization was made. The next scale of analysis, *88-sections*, reveals additional patterns ; for example, a suburbanized area of high status appears along Mahim Bay and Matunga. The most detailed study, *437-circles*, indicates that the high status areas downtown are commercial. A few high SES areas are shown scattered even further towards the edge of the city as small suburbanized colonies, but the very lowest status areas dominate the outermost fringes of the city. The industrial and port areas, established for a long time, do not have such low status ratings as the more distant areas, revealing that long-time industrial urbanization has its socio-economic advantages.

Familism : male migrant areas

Data relating to family structure were available for only four cities (Bombay at each scale of analysis), but on each a "familism" index appeared. However, the factor structure reveals that this element of India's urban ecology is not a measurement of degree of family specialization according to stage in life cycle, as in the United States, but one which picks out areas in which vast numbers of men who are working in the industries of the city come to live. They live without women, though they may be married to women they have left in the village, at least temporarily. A similar factor was noted for Cairo, Egypt by Abu-Lughod (1968b). Statistically independent of SES, some of these neighbourhoods of factory workers may approach middle-class status, or else the factorial output indicates that many of the migrants live in neighbourhoods which are predominantly middle class rather than lower class.

The maps (not reproduced here) of the factor scores indicate distinct land use associations. In Bombay, the fifteen-ward study shows houseless male workers living without their families to be concentrated in the dock areas, the industrial areas, and downtown. Much as one would expect, the houseless workers gravitate to the areas where jobs for unskilled workers are available. Conversely, the most marked positive area of family life is on the fringe of the region where, one presumes, land is available for home building, and indeed, the area may not yet be seen as fully urbanized.

The 88-sections study picks up the same areas again in finer detail, revealing additional areas of family dwelling without houseless street-sleepers. Unlike the downtown areas, these areas do not attract transient labourers for, despite their affluence, they do not have many jobs to offer. The fine grained 437-circle analysis come closer to pinpointing the areas of job availability for transient male labour, separating them from areas of stable residence without commercial land use. Thus, harbour and textile mill areas are strongly marked as low on the familism index ; suburban family areas rank high.

Kanpur exhibits a similar separation of industrial areas from residential areas on this index. Interestingly, Kanpur is a city surrounded to the northwest and the southwest by industry. The northwestern quadrant, however, was pioneered more by the British and adjoined the cantonment, no exaggerated familism index, high or low, is in evidence. This may result from British policies in "their" areas of the city, or from an overlap within the same census ward of residential and factory areas so that one balances out the other. The area to the south of the city includes a number of new industries, as well as older ones, and the main railway depot for industry. It has a decidedly low familism index (a factor score of —5.92). The areas especially high on familism seem to be agricultural areas on the fringe of Kanpur, areas quite removed from industrialization.

The familism index in Ahmedabad appears in two sets of factor loadings. As factor VI it picks out the railway depot ("Railwaypura" ward ; factor score—4.53) and the job-rich, highly transient area around it. The workers here are not in industry, but rather in the commercial and transport occupations connected with the depot. As factor VII, it distinctly picks out areas of males in household industry and in construction, particularly the new area developing across the Sabarmati river. Similarly, it picks out zones of household industry for men, where hand weaving and some handicrafts are still practiced in the southern areas of the city both within and without the walls. The areas highest on the familism index are : Ranip, a rural area recently annexed to Ahmedabad and so far distant that is not yet mapped ; the military cantonment which houses an elite ; and ward 22, Gomptipur, which was for some time a separate town, suburban to Ahmedabad.

Finally, the Madras familism map picks out most clearly the harbour area, the natural destination of men travelling alone and seeking work.

The most significant aspect of the communal element of India's urban ecology may well be the fact that Government of India refuses to collect data on caste and communal factors in the official census. The only data available on communal affiliation, therefore, are in the surveys of Poona and Sholapur.

In the study of Poona a clear pattern of communal residence appears. Muslims are not prevalent in neighbourhoods where Brahmins predominate, and *vice versa*. Similarly, Brahmins and scheduled caste members do not in general share the same localities. Indeed, in Poona, the Brahmins are composed of several groups, most prominently of Konkani Brahmins originally from the Konkan coast of western India, in the Ratnagiri District of modern Maharashtra State, and of Deshasta Brahmins who come from the Desh or the "country", the internal plateau regions of Maharashtra (Karve 1963). In 1937 the Konkani

Brahmins were heavily over-represented in Sadashiv ward while the Deshastas were more prominent in Budhwar, Shukrawar, and Shanwar. Konkanis are also more highly represented in the highest income brackets and in the highest levels of occupational types. Within an already elite group, the Konkanis (or Chitpavans as they are often called) are the most elite.* A similar separation between the two principal Brahmin groups appears for 1954.

The Sholapur survey shows that wards with high proportions of Brahmins are also high in Marathas, the dominant indigenous caste of the region in terms of numbers. While the Maratha caste is not high in socio-economic status, working essentially in farming in the rural areas and in craft and industrial jobs in the cities, it is quite respectable in ritual status. The Marathas are the traditional rulers who employed the Brahmins as the administrators of the empire. Conversely, Lingayats and non-Marathis tend to be disproportionately high in the same wards, again no surprise on ritual lines since neither has a niche in the local Hindu caste structure, non-Marathis since they are immigrants and Lingayats because they follow a deviant form of Hinduism. Occupationally, however, they bridge a wide spectrum of activities from large scale trade to factory workers and private artisans.

The Sholapur study is unique in that it shows communal status rather than socio-economic status as the most significant variable in residence patterns, Sholapur did not show the clear presence of a socio-economic status variable. And though Factor I, the communal status factor, here is linked with socio-economic status, this factor is recessive while the caste factor is dominant. The analysis of Sholapur broke into nine factors, highest of any city, and it had 33 variables for consideration. This finer-grained information base may well be responsible for the greater weighting of the communal factors as compared with the SES variables, and clearly the two are interrelated. Sholapur remains as testimony of the power of caste and communal factors in the

*Simply in terms of residence choices, this division in the Brahmin community reinforces the model of different groups sorting themselves out residentially according to SES. Dynamically, the preeminence of the Chitpavans may well owe to the fact that the rulers of Poona in its first heyday, the second half of the eighteenth century to 1818, were Chitpavan Brahmins. Doubtless they gave special encouragement to fellow caste members.

formation of neighbourhood in a modern Indian industrial city.

Clustering of communal residential areas is, of course, predictable, and in Poona reveals remarkable long-term stability. Over a period of 150 years, from the ruler's social survey of 1822 to the Surveys of the Gokhale Institute in 1937 and 1954, the centre of the old city retained its social pattern quite apart from an almost tenfold gain in population. The Brahmin high-caste high SES group dominated the western region of the city between the Mutha river and the present-day Katraj aqueduct. The centre of the town, along a north-south axis, was and is dominated by the market functions of the town and houses the traders, the Muslims, and the non-local groups of the society. The eastern and north-eastern quadrants of the city house the untouchables and scheduled caste members for the most part. Thus the old central city retained a clear caste-class correlation over time, and residential patterns clearly separate out along this line. However, Poona has recently experienced suburbanization, and for the first time large numbers of people are moving away from the centre of town. In the suburbs, class and occupation are bringing together groups from diverse castes. In addition, newer migrants have tended to cloud the older outlines.

VARIATIONS IN TIME AND AREA

The Poona and Bombay materials provides further insights into variations of the results in time and area.

Data on Poona are unusually rich both in quantity and time. Materials are available from the records of the Maratha administrators which date back to 1822, and from the surveys of 1937 and 1954. This facilitates examination of materials on the size of the city, its density, the external connections with other cities, the economic functions of Poona, and the social areas of the city over time, and provides strong indication of the effects of various processes over time.

Poona began as a temple and fort complex, located slightly away from the river bank, and it had in the early seventeenth century many of the features of the classical Indian city. The priests and the administrators, grouped themselves around the temple and fort-administrative area. Off to the east, on marshy land undesired by other groups, the servants and untouchables resided.

Rivers and streams flowing from south to north divided the city into three major sectors. In the west between, the Mutha river and now a dry stream which followed the course of the contemporary Katraj aqueduct was the Brahmin sector of the city. In the central city, between the old river and the Nagzari stream was the bazaar and shop area of the city inhabited by commercial and artisan castes and by non-indigenous people. Initially the untouchables lived in the eastern sector.

Poona took on great importance in the eighteenth century as the location of the court of the local ruler and also of the *Peshwa* or prime minister and *de facto* ruler of the spreading Maratha Confederacy. The city expanded during this time, essentially in a north-south direction, following the earlier three divisions and natural topographical features. Successive areas were added to the city by the Peshwa, who turned each new ward over to a nobleman for management and ordered him to bring a full population to it. Sometimes people were already living there, sometimes they came later ; the dates given to wards indicate the year of assignment to the nobleman. However, as wards were added, their residents took on the same character as earlier ones in their section, and so the 1822 analysis revealed clear communal distinctions as the prevailing element in the town's social geography.

Expansion of the town beyond the triangular area bounded by the hilly ridges to the south, the river to the west, and the marshy land to the east took place only in the nineteenth century, and more prominently in the twentieth. The British after conquering Poona in 1818, built their cantonments to the east and the north, Poona Cantonment (the civilian area) and Kirkee Cantonment (the military area), respectively. Later in the nineteenth century, Indian settlement spread west of the river, establishing there an outpost of the old Brahmin sector in a highly intellectual area of colleges and schools. Finally, the twentieth century, especially after independence, has seen the development of Poona as an industrial area with the spread of factories along the railway lines, especially to the west, toward Bombay. Both the 1937 and 1954 analyses reveal the additional British and industrial ingredients in the town's social map.

The physical expansion of the city, especially in recent years, has not quite led to suburbanization. The population has continued to pour into the old core city, boosting the population densities there. Maps from Gadgil's study (1945, 1952) indicate the increasing density of the central area. A later compilation by Brush indicates that densities are still increasing in the central areas, and the relative sparseness of population in the surrounding area. Brush notes "in 1881 Poona's population was recorded as 99,421 ; fifty years later it had increased to 213,680, with little change of area. By 1961 the old city had been enlarged more than ten times, from 2,390 acres, to 27,190 acres and the population had grown to 597,419, but without any large shift of people into areas incorporated into Greater Poona" (Brush 1968). Brush correctly suggests that if Poona follows patterns of western cities, with the spread of cheap and rapid public transportation in the city, the suburban exodus will soon come. At present workers travel by bike from the central city to jobs on the periphery. The city now seems ready for the jump to a reverse situation.

The size and density changes have come about largely by a change in the economic, political, and social functions which Poona has played throughout the historical period covered. It changed from a governmental centre under the Marathas to a British governmental and military centre, the headquarters of the southern command of the Indian army, to an industrialized metropolis to some extent before Independence and much more rapidly since then.

What ecological change has been associated with these functional and demographic changes ? At least until the last decade, they did not upset the basic structure of the core city. But the creation of the cantonment areas by the British did lend to these new areas quite different characteristics from the older parts. There was greater diversity, with more foreigners and fewer Hindus. Further change was introduced by industrialization and by the consequences of partition. Each factory attracted its cluster of migrant workers, and the general acceleration of rural-to-urban migration impelled many lower caste Indians to the cities, where they filled most of the empty spaces, clouding the earlier social differentiation. When more data from the 1961 census become available, they will, therefore, show substantial smoothing of the social lines that remained essentially stable in the three analyses, 1822-1954.

Bombay's three-level analyses provided the opportunity for exploring the effects of differing geographical units of observation on the results of factorial ecologies of Indian cities. By and large, the factorial results were remarkably stable, which is pleasing. However, as the finer-grained analyses were undertaken, much greater insight was provided into the city's social map. With 437 units of observation, for example, docks were clearly separated from wealthier higher socio-economic areas behind them, a feature clouded by the 88 sections and 15 wards. What the results, therefore, indicate is a clear need in factorial ecologies for some notion of the size and character of neighbourhood groupings in Indian cities ; hopefully this will serve as a guide to appropriate observational units that will provide a proper degree of geographic detail to general ecological themes repeating themselves through a variety of scale of analysis.

CONCLUSIONS

To the extent that the evidence analyzed reveals, socio-economic status appears to be the dominant of these over arching themes in the residential geography of the Indian city today. The communal and caste status of the classical texts and of village India, and of Poona in 1822, is being transformed into class status as an outgrowth of city life (Rosen 1965). While this is taking place, residual caste status and class status reinforce each other, so that the dominant spatial pattern remains that of high status neighbourhoods in central area and low-status neighbourhoods at the periphery.

If ecology is to be understood as process in time, much more historical study of urban ecology is needed in India where traditional patterns have great longevity and where social areas display great resistance to change. Among the active forces which have introduced new elements into the social geography of the Indian city are : the British cantonments, built on different social bases than old Indian areas, often creating a dual city structure and a second set of higher socio-economic status central and lower socio-economic status peripheral neighbourhoods ; newer institutional areas devoted to government or education, although thelatter tend to be Brahmin and the former (even in Chandigarh) tend to repeat the traditional core-to-periphery distribution by

SES ; and industry. It is this latter element that created a second factor in India's urban ecology by 1961 : the family-type distinction between familial areas and zones housing new male migrants, clustered around the commercial core of the city, docks, mills, and new factory estates. Outlying industry, rather than improved transportation, has also promoted limited "suburbanization" and is one of the contributing causes of increasingly specialized neighbourhoods in all cities.

What seems clear is that the increasingly diverse bases of social and economic power, which city life in modern India is generating, are transforming the urban structure. No longer are land or rulership or priesthood be bases of status or neighbourhood and community, although their effects are still marked in India's urban landscapes. The issue is whether the emerging forms are converging on the model of the industrial metropolis, as suggested by Berry and Rees in Calcutta, or whether some new synthesis of traditional and modern will emerge.

SIX

JOHN E. BRUSH

RECENT CHANGES IN ECOLOGICAL PATTERNS OF METROPOLITAN BOMBAY AND DELHI

THE 1971 Census of India yields evidence of significant shifts in the patterns of growth and demographic structure of metropolitan Bombay and Delhi. There are two features: first, a small beginning of population deconcentration in the central areas; and second, the emergence of sectorial grouping of population by social status towards the peripheries.

Prior to 1961 most Indian cities, including Bombay and Delhi, had absorbed much population through crowding more and more people into areas already densely occupied. The result in most of the cities of India was progressive increase of central density, in some cases to levels twice what it had been 50 years previously. Such a trend is contrary to the urban spatial growth patterns in Western Europe and North America where, in the past several decades, densities have increased in the intermediate and peripheral zones of urban areas while decreasing in the central zones (Brush 1970). Berry, Simmons and Tennant (1963) observed this spatial growth pattern in India and took it, along with other data, as evidence of divergence in the spatial structure of change in non-Western cities, as compared with the structure of Western cities. Elsewhere (Brush 1968) four different types of urban spatial structure existing in India were identified, dependent upon the economic and political circumstances of city formation and associated population distribution. However, on the basis of 1961 data it was anticipated (Brush 1973) that the patterns of Indian cities would ultimately conform to a general model of central decline and centrifugal population growth patterns.

POPULATION GROWTH AND DENSITY CHANGES IN BOMBAY AND DELHI

It is seen (Table 6.1) that among the three national metropolises of India, Bombay experienced the largest absolute population increase and Delhi the largest relative increase. The population of the Calcutta metropolitan area increased less than 23 per cent from 1961 to 1971 continuing a trend of relatively sluggish and diminishing growth in the previous two decades, while Bombay and Delhi maintained higher growth rates of about 44 and 55 per cent respectively. Preliminary analysis of population distribution in Calcutta from the 1971 Census (Brush 1976) indicates that peak density in central Calcutta had risen from 1961 to 1971 contrary to the patterns in Bombay and Delhi. In this paper, only the latter two urban areas have been selected for analysis of geographical patterns of growth. It should be noted that all concentrations are expressed in terms of gross residential density

TABLE 6.1 : **Population growth in India's three national metropolises, 1961—1971**

Metropolis	Population 1971	Population increase 1961—71	Per cent increase 1961—71
Calcutta Urban Agglomeration	7,031,382	1,294,685	22.57
Greater Bombay	5,970,575	1,818,519	43.80
Bombay Island	277,933	298,445	10.77
Bombay Suburbs	1,380,113	1,520,074	110.14
Delhi Urban Agglomeration	3,647,023	1,287,613	54.57
Old Delhi (former Walled City)	388,648	24,425	6.28
New Delhi Municipal Corporation	261,545	40,256	15.39
Delhi Cantonment	36,105	21,234	58.81
Delhi Municipal Corporation excluding Old Delhi	1,673,110	1,201,810	71.83

by census statistical divisions. The area measurements on which these calculations are based were taken from official reports for Bombay but in Delhi, where official data are lacking, the census divisions were plotted on a base map on the scale of 1 : 20,000 and areas estimated by the author. Other data used in this study were taken from district census handbooks of Calcutta, Bombay and Delhi for the years 1961 and 1971.

Greater Bombay has an area of 603 sq km with a general density of 99 persons per hectare in 1971. It comprises Bombay Island with an area of 6,777 hectares and a density of 453 persons per hectare and Bombay Suburbs on Salsette and Trombay Islands with 53,523 hectares and a density of 54 persons per hectare. Highest densities occurred on Bombay Island, where two of every three residents were found in 1961.

In 1961 the maximum concentration on Bombay Island was 3,756 persons per hectare, the highest peak of urban density recorded in India, representing so-to-speak an Indian demographic "Mt. Everest". The core of high density (over 500 per hectare) is located in the mixed residential-commercial interior part of the Island, close to the docks and railway terminals with extensions northward in the old cotton mill district and the Mahim river causeways (Fig. 6.1). In suburban Bombay on Salsette Island densities rarely exceed 250 per hectare and diminish to less than 50 per hectare towards the periphery.

The peak density area with more than 2,000 persons per hectare is located in the heart of Bombay in and around Bhuleshwar, Kumbharwada, and Kamathipur, which the British once called the "Native City" with its congested *bazar* streets, pedestrian alleys, interior courts and overcrowded *chawls*. Concentration is much less in the central business district and main governmental administrative area, centered around Hutatma Chauk (Flora Fountain) in "The Fort", which was the site of first British settlement (Fig. 6.2, after Rajagopalan 1963).

Looking at the percentage change of population from 1961 to 1971, it is notable that decreases occurred both in the central commercial-residential area, where the peak density in the Bhuleshwar Division dropped about one per cent to 3,712 per hectare and in the office and finance areas, particularly in the North Fort Division, which declined 23 per cent (Fig. 6.3). Decreases are observed also in certain residential areas fronting

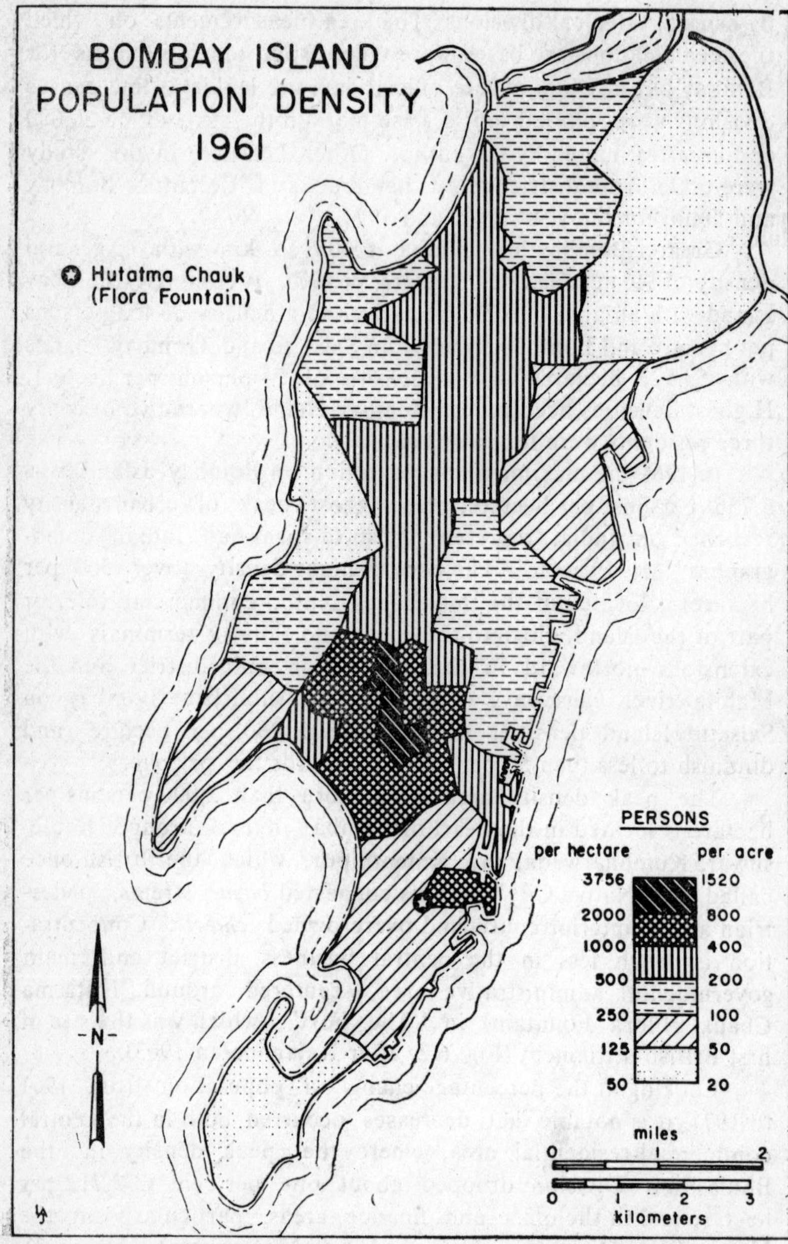

FIG. 6.1 : *Bombay Island population density, 1961*

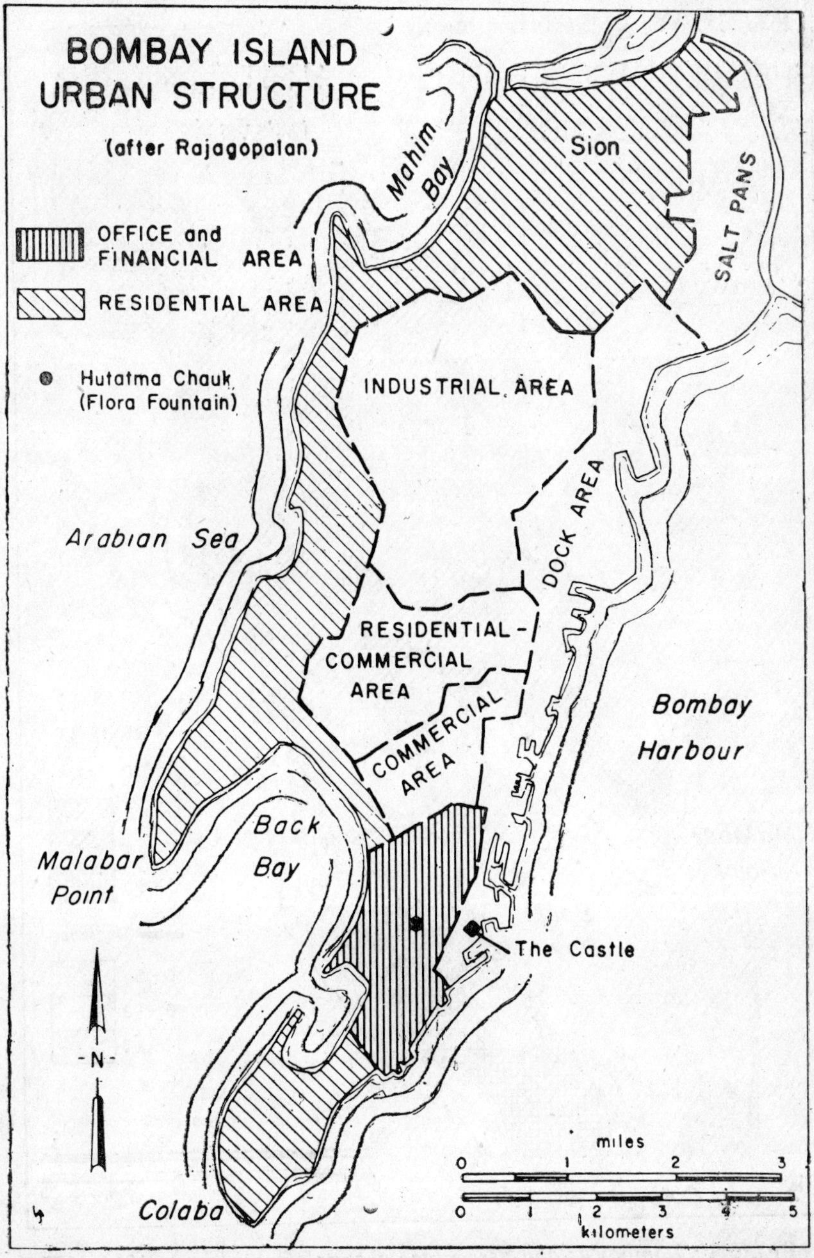

FIG. 6.2 : *Bombay Island urban structure*

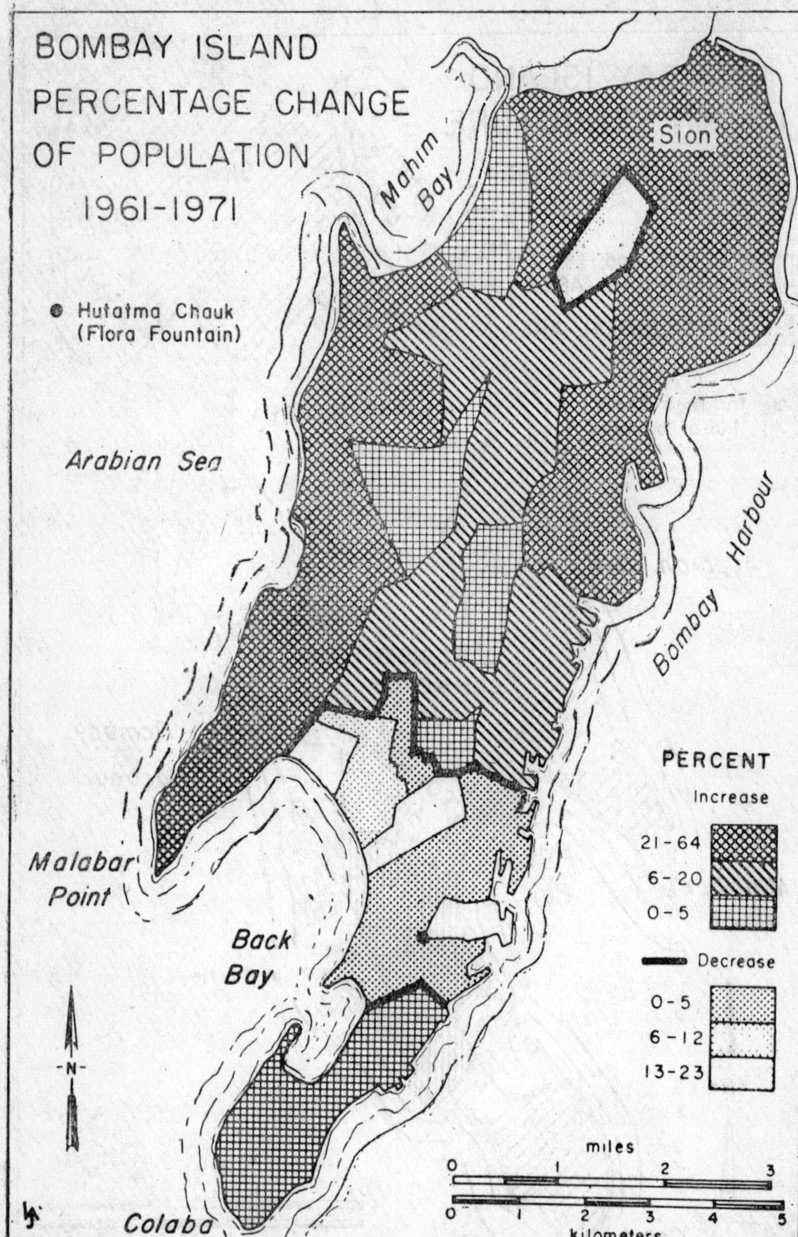

FIG. 6.3 : *Bombay Island percentage change of population, 1961-1971*

on Back Bay and on the north Matunga Division. The greatest percentage of growth on the Island occurred in residential areas facing the Arabian Sea and in the peripheral areas in and around Sion to the north. The net change of population on Bombay Island between 1961 and 1971 amounted to an increase of less than 11 per cent, while there was ten times as much growth in the Suburbs on Salsette and Trombay during the same time period. By 1971 almost half the total population of Greater Bombay, some 2,900,000 persons, lived in the suburbs. Thus, it becomes clear that Greater Bombay has entered a period of rapid peripheral growth, while the Island is approaching population saturation and may be expected to stabilize or decline in the future.

The Delhi Urban Agglomeration (or metropolitan area), as defined in 1971, has an area of 446.26 square kilometres with general density of about 82 persons per hectare. In comprises the Delhi Municipal Corporation with 36,055 hectares in 1971 and density of 91 per hectare, New Delhi Municipality with 4,274 hectares and density of 71 per hectare, and Delhi Cantonment with 4,297 hectares and density of 13 per hectare. The average density is approximately 800 persons per hectare in Old Delhi, the former walled city of Mughal times, rising to a maximum of 1,663 per hectare in one census division. Almost equal densities are observed in extensions "outside the wall" towards the west, but elsewhere the concentration of population declines abruptly. Outlying belts of moderate density occur in the southern sector in and beyond the limits of the New Delhi enclave and along the main routes of transportation to the east, north and west of Old Delhi. The low density in Delhi Cantonment in the extreme southwest is due to restrictions against non-military residential land use and the presence of Delhi International Airport (Fig. 6.4).

The pattern of growth in Delhi between 1961 and 1971 was clearly centrifugal. Note that population in Old Delhi gained only 6 per cent, while the remainder of the Delhi Municipal Corporation gained over 70 per cent. It must be explained that the total area of the corporation was increased about 50 per cent by annexation of 40 peripheral villages during the decade. Most of the 1.2 million added during the decade, however, was either natural increase or in-migration absorbed in the areas of

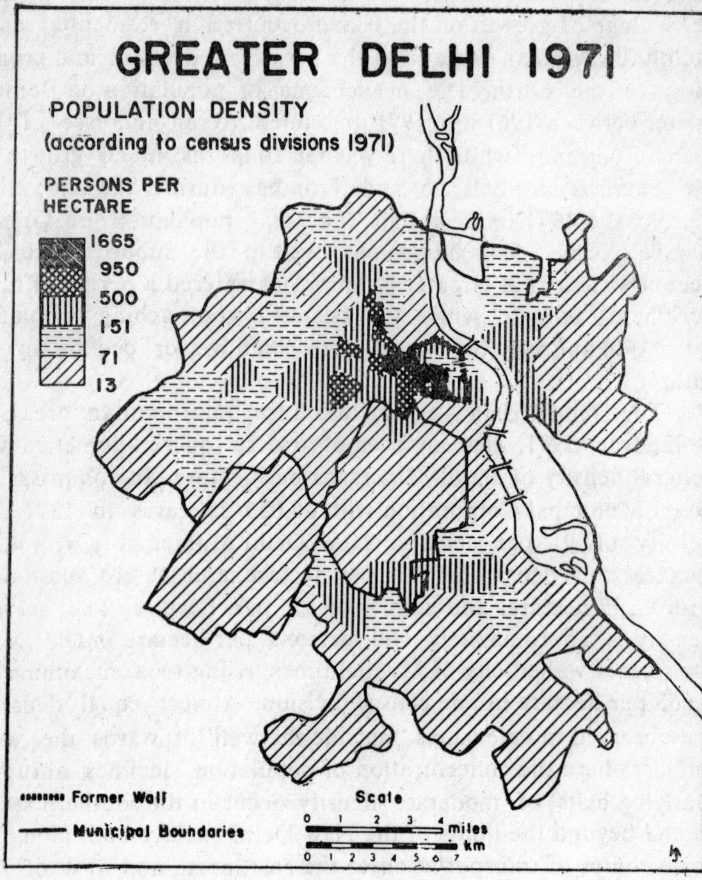

FIG. 6.4 : *Greater Delhi, 1971*

intermediate density, located outside the pre-existing central concentration.

It is not possible to compare population distribution in 1971 and 1961 in Greater Delhi with precision because of alteration in the boundaries of census divisions, except in Old Delhi and New Delhi. In the first instance, Old Delhi, the area of population decline is situated around the Delhi Railway Station towards the Red Fort and in the congested commercial-residential area south of Chandni Chauk. It seems evident that centrifugal tendencies are overcoming the centripetal forces of the past, perhaps associated with increases of commercial and other non-residential land uses around the nodal point of the old city—Chandni Chowk and the Municipal Building. In the second instance, New Delhi, population decrease has occurred in the principal public areas, located on both sides of Raj Path between the Secretariat and India Gate from Jan Path on the north to Lodi Road on the south. Population was not, and is not, concentrated here. The density varies between 30 and 50 per hectare and the relatively slight decrease would seem to have been caused by displacement of residential uses by governmental administrative functions and by retail shopping, banking and other commercial activities around Connaught Place and along the approaching business streets.

The evidence adduced above shows that recent changes in population distribution in metropolitan Bombay and Delhi conform to the general model of the spatial growth pattern of the more urbanized, economically developed parts of the world. The results of factor analysis of the 1971 census data in the two metropolitan areas suggest that the emerging socio-economic structure also parallels the patterns observed in some Western countries.

SOCIO-ECONOMIC STRUCTURE : BOMBAY

The central areas of most cities in India, especially those of indigenous origin, retain a locational advantage for business activity and a certain traditional prestige as the place of residence for people of high social and economic status. Here were found the families of the old aristocracy, the land owners, wealthy merchants and bankers, and generally the people of high ascriptive ranking—the upper castes. People of lesser privilege, little wealth and low social rank, particularly the Scheduled Castes, often find themselves forced to live on the edges of the city. Such a spatial structure was common in India before the British arrived (Brush 1962) and survived into the mid-20th century in the typical Indian city. The pre-modern pattern persisted under the colonial rule but today the traditional urban elite find their power and prestige being supplanted by a new economic and social class in which membership is based on education and on access to managerial, technical, and administrative positions in industry, government and professional services.

The factor-ecologic structures of Indian cities, identified from 1961 and earlier statistics, have shown both zonal patterns with the population of high social rank in the central location and sectorial patterns in which population of high achieved status have concentrated towards the periphery (Berry, *et al.* 1969, 1971 ; Singh 1975). Thus, it appears that India's urban development is in transition, not strictly "pre-industrial" in Sjoberg's terms (1960) yet not exhibiting fully the features of modern industrial social ecology. The results of factor analysis of 1971 census data on Bombay and Delhi presented in this study identify areas where people of the new elite class have concentrated in largest numbers during the period of rapid urban growth since Indian Independence was attained.

The five factors revealed in Bombay, which together account for about 74 per cent of the variance occurring among the 88 census divisions are examined here (Table 6.2). Factor 1 shows strong positive association of high literacy for both males and females with concentration of workers in trade and commerce and in services. There is also a weak association of gross population density and per cent of females employed in Factor 1, while workers in farming or in livestock rearing, fishing, and so on, and munufacturing show moderate to weak negative relationships to the foregoing variables. This factor, which explains about 20 per cent of total variance, can be considered as an indicator of high achieved social and economic status. Factor 2, on the other hand, is an indicator of low social rank, showing strong positive association of scheduled caste population with workers in manufacturing and in transportation and communication. Percentages of females working and workers in agriculture show moderate degrees of negative association in Factor 2. This factor explains 14 per cent of the total variance in Bombay and is almost equalled by each of the next two factors. Factor 3 may be described as a measure of high housing concentration, expressed as persons/house and persons/household, positively associated with the percentage of females working, but negatively associated with concentrations of workers in manufacturing and construction industries. Factor 4 shows positive relationships between density of population, and concentrations in household industry and trade

TABLE 6.2 : **Factor analysis—Bombay 1971 (Rotated factor loadings over + or —.300)**

Variables	Factor 1	Factor 2	Factor 3	Factor 4	Factor 5
Persons per hectare	+.329	—	—	+.724	—
Persons per house	—	—	+.867	—	—
Persons per household	—	—	+.877	—	—
Ratio of females to males	—	—	—	—	+.932
% of population : scheduled castes	—	+.657	+.335	—	—
Female/male ratio : scheduled castes	—	+.665	—	—	—
Literacy : % of males literate	+.915	—	—	—	+.382
Literacy : % of females literate	+.936	—	—	—	—
Total males : % workers	—	—	—	—	—.911
Total females : % workers	+.316	—.568	+.367	—.673	—
Total workers : % in farming and farm labour	—.569	—.626	—	—	—
Total workers : % in livestock, forestry, fishing, and other primary production	—.586	—	—	—	—
Total workers : % in household industries	—	—	—	+.756	—
Total workers : % in manufacturing industries	—.324	+.731	—.352	—	—
Total workers : % in construction	—	—	—.527	—	—
Total workers : % in trade and commerce	+.652	—	—	+.558	—
Total workers : % in transport and communication	—	+.522	—	—	—
Total workers : % in services	+.615	—	—	—.606	—
% variance explained	20.5	14.5	13.6	13.2	11.9

and commerce, while negative relationships exist with percentage of females working and concentration in services. Factor 5, which explains 12 per cent of the total variance, shows that a strong concentration of male workers is negatively related to the ratio of females to males in the population as a whole and in Scheduled Caste population.

Examination of the geographical distribution of each of the five factors brings out some basic features in the emergent urban structure of Bombay. The highest scores on Factor 1 are found at Colaba and the South Fort and in a belt from Crawford Market to Girgaun on Back Bay and Malabar Hill, although there are two small areas with high scores towards the north end of the Island. Another area of upper class population occurs in the belt along the western side of Salsette Island with maxima in the extreme southwest near the Mahim river causeways connecting Salsette to Bombay proper. The divisions with low social rank show the extreme concentration on Bombay Island in two areas : one, near the docks along the Harbour side and the second, on the west side from Parel and Chinchpokli northward to the Mahim river. Generally, population concentrations of low social rank extend through the middle of Bombay Island, across, Trombay and over the south-central and eastern parts of Salsette up to Mulund on the Greater Bombay Corporation limits.

Factor 3, which measures housing concentration, shows the highest scores, not surprisingly, in two well-defined areas around the Fort and north central Bombay, beginning in the most congested residential-commercial and industrial areas. Bombay Island contains the vast majority of census divisions with positive scores in Factor 3. There are, however, a few areas of housing concentration on the west side of Salsette. Note in particular the "islands" of Madh and Manori, small coastal villages on the Arabian Sea which have only roundabout land connections to the main suburban belt of western Salsette. Factor 4, which indicates correlation of high population density with household industry, or trade and commerce, shows highest scores within the area of density in excess of 1,000 per hectare. Around this cluster in central Bombay there is an area with scores above zero, but in suburban Bombay the distribution is spotty. Factor 5, male worker concentration, shows highest scores in and around

Colaba, the Fort, the Crawford Market area, the dockyards and cotton mills. Scores exceed zero in almost the entire area of Bombay Island, except for a few census divisions in the heart of the city and at the northern end. On Salsette Island scores above zero tend to follow the inland hill margins or the coast.

SPATIAL STRUCTURE : BOMBAY

The spatial structure of socio-economic classes in Greater Bombay has been delineated by combining Factors 1 and 2 (Fig. 6.5). Since the factor scores are inversely related, the map clearly differentiates areas of high social status with positive scores in Factor 1 and negative scores in Factor 2 (1+ and 2—) from areas of low social rank with positive scores in Factor 2 and negative scores in Factor 1 (2+ and 1—). The two other combinations are either areas with scores in both factors (2— and 1—). The former represent areas of transition between high social status and low social rank with positive components of both while the latter represent areas with components of low social status and workers in agriculture, fishing and other kinds of primary production, which occur as negative features in Factors 1 and 2.

Certain aspects of socio-economic structure revealed by Fig. 6.5 are independently corroborated by Rajagopalan (1963) :

> . . . the existing ecological structure of the City (Bombay Island) may be described as a series of successive zones . . . . The geographical centre of the City is occupied by the industrial zone. In the north this zone gradually merges into the working class residential area which in its turn is succeeded more abruptly by a better class residential area. Immediately south of the industrial zone is located the residential-cum-commercial zone. Between these two zones intervenes a zone characterized by physical deterioration and social disorganization and chiefly inhabited by the poorer sections. As we proceed further south, the residential-cum-commercial zone gradually turns into a business and trade centre, where residences are relatively few. Further beyond, in the extreme south is located the core of the City (the Fort), and where the bulk of the white-collar workers are employed . . . the Back Bay in the south and the Malabar

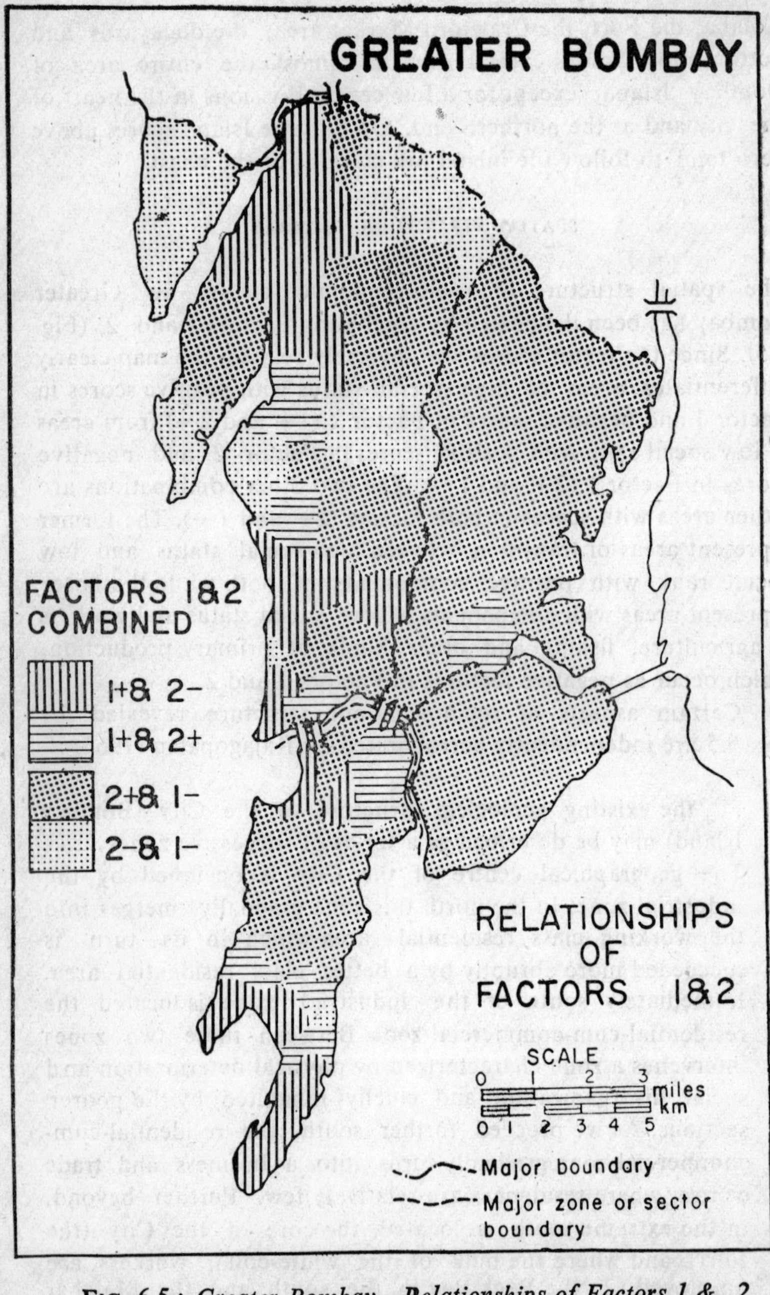

FIG. 6.5 : *Greater Bombay—Relationships of Factors 1 & 2*

and Cumballa Hills in the south-western tip contain the wealthy residential areas.

Comparison of Fig. 6.5 and Fig. 6.2 (Rajagopalan's map of Bombay Island) confirms existence of the principal area of low social rank in the geographical centre of the Island and the Harbour side on the east, where it extends along the entire length of the dockyards. To the north there is an abrupt juxtaposition of upper-class areas and to the south a zone of transition to the main upper-class residential area on the island, fronting on Back Bay and extending southwest to the Malabar Hills. In the south another zone of transition bounds the Fort with another upper-class residential area extending to Colaba.

Rajagopalan sought confirmation of Burgess' zonal model of urban ecological structure, derived from observations in Chicago, by identifying the central industrial area of Bombay as the centre for the concentric zones which he observed. If he had taken the Fort, which he recognized to be the "core of the City", as the central zone he would not have found any support for Burgess' hypothesis. Burgess regarded the central business district of Chicago as (1) the central zone, followed by concentric rings of, (2) manufacturing industries and changing land use, (3) social disorganization, physical deterioration and residential succession, (4) Working-class residences, (5) middle-class residences and (6) the residences of upper-class commuters living on the city periphery and in the suburbs. The spatial arrangement of social zones is reversed in Bombay, with the high-class residential zone and a mixed residential-commercial area and highest population density located within the first 3 to 4 kilometres, followed by a zone of transition and labouring-class slums in the industrial and dockyard zone from 2.0 or 3.5 to 7 kilometres from the city centre (Fig. 6.6). Beyond this lower-class zone are certain upper-class residential areas located between 7 and 10 kilometres from the city centre. These northernmost upper-class areas in Mahim, Matunga and Sion are linked tenuously to the prime residential area of Malabar and Cumballa Hills by a narrow residential strip with upper-class population along the Worli sea face, as shown on Rajagopalan's map. The map gives no evidence of this linkage

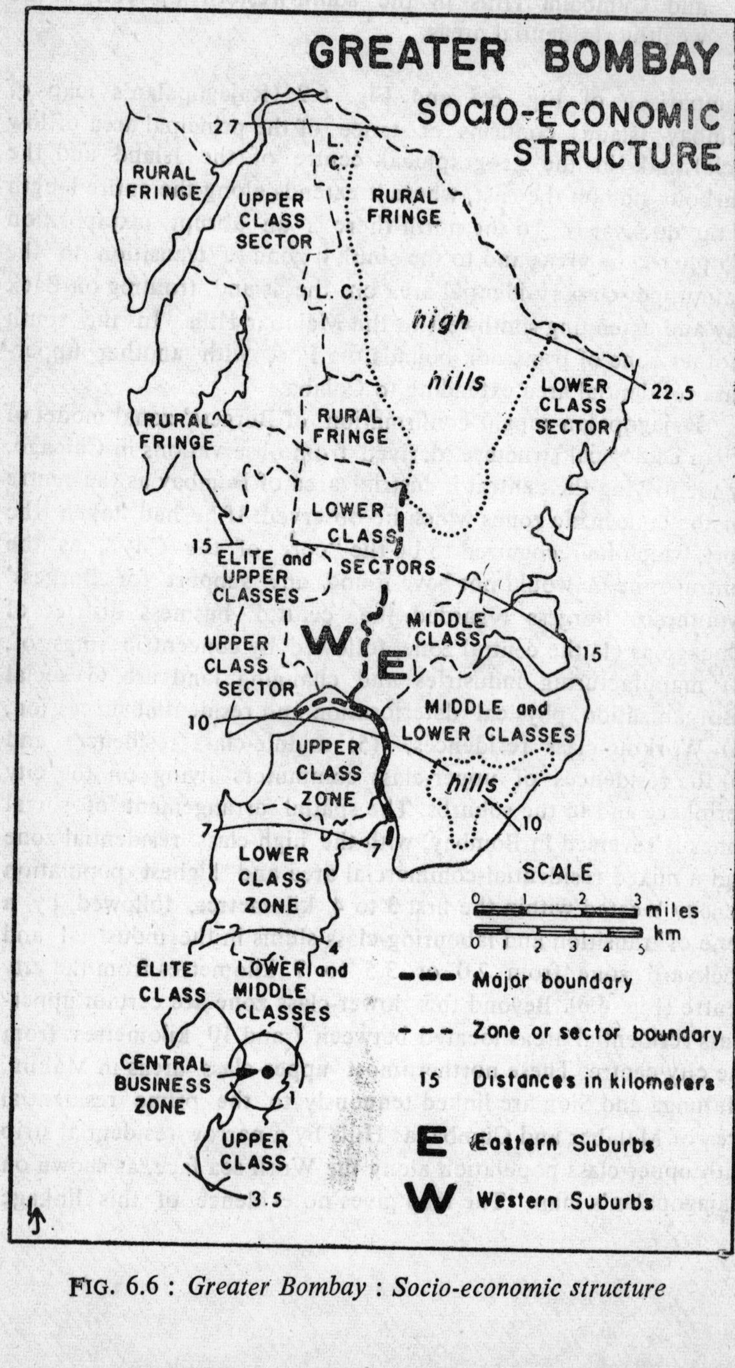

FIG. 6.6 : *Greater Bombay : Socio-economic structure*

through Worli and Love Grove, evidently because the population of high social status is insignificant in comparison with the larger population of low social rank found farther inland in the same census division. Corroboration of these basic socio-economic features of Bombay Island is found in Nissel's statistical study (1977) of the functional structure of Bombay, based on the 1961 census data. The location of the upper-, middle-, and lower-class residential areas, as well as the principal commercial and industrial zones thus is independently confirmed.

Thus, factor analysis reveals that the spatial structure of Bombay proper is predominantly zonal, but unlike the Burgess model, the residential zone which is closest to the city centre attracts upper-class population, while labourers and the lower social classes are found near the dockyards and concentrated in the relatively distant industrial zone at the geographical centre of the island. Decreases of population already noted, which occurred during the last census decade in areas up to 2.5 kilometres from the city centre, suggest that residential decline and displacement by business functions has begun to affect the Fort and the adjacent mixed commercial area of maximum density. Of course, this is the typical process of change observed for many decades in European and American cities. However, in the old elite residential area of Malabar and Cumballa Hills, located about 3 kilometres from the city centre, population gains are recorded and much new high-rise apartment building is occurring. Population remains essentially fixed with some areal expansion occurring in Colaba on land reclaimed from Back Bay, an area situated within 2 to 3 kilometres of the central business district. The socio-economic structure of Bombay Island in 1971 continued to be compact with much of the upper-class population living relatively close to the city centre. The residential concentration of Bombay's "ruling elite" in a strip extending from the peninsula of Colaba along the shores of Back Bay to Malabar and Cumballa, subjectively described by Patel (1974, pp. 313-14), verifies the identification of this middle- and upper-class area through factor analysis and shows the degree to which the historical social ecology developed in Bombay during the late 19th century has persisted (Kosambi 1980, pp. 121-135).

In view of excessive congestion in Bombay Island, it is not surprising that accelerated population growth has been absorbed in the last decade on the adjacent islands of Salsette and Trombay. There are two main rail lines on Salsette with parallel highways, respectively, the Western Railway and the main road to Gujarat on the western side and, on the east side, the Central Railway and the main road through Thana, linking Bombay to central and northern India. Heavily forested hills form a triangular area between these two transport corridors and there are extensive mud flats and tidal channels bordering the creeks draining into the Arabian Sea and Thana Creek on the east. New express highways have been built parallel to the old main roads. But no other overland route of access to Bombay Island existed until 1971 when the Thana Creek Bridge was built linking Trombay to the mainland. Salsette and Trombay comprise the new urban zone of Greater Bombay, although industrial development on the mainland shore has begun and a planned city of "New Bombay" is projected on the east side of Thana Creek and Bombay Harbour.

The shift to a sectoral pattern is a most striking fact about the spatial structure of suburban Greater Bombay, in contrast to the zonal pattern of the city proper. The predominantly upper-class sector of western Salsette is clearly differentiated from the predominantly lower-class sector of eastern Salsette and Trombay on my map. Although Rajagopalan does not treat the suburbs, this pattern is elucidated in part by the observations of Dixit (1959) and Deshpande (1973). In the eastern sector on Salsette Island there is a ribbon of factories, extending from Kurla northward along the original main road to Thana on the higher and more accessible ground on the west side of the rail line. On this employment base both planned and unplanned housing tracts have sprung up to fill the spaces between and around the factories. Almost none of the dwellings are of good quality and they are occupied by workers of predominantly low social rank with the exception of Ghatkopar, an area which is more attractive to upper-class residents. The western side of Trombay Island has become heavily industrialized with oil refineries and petrochemical plants and is predominantly low in social rank, despite some middle-class housing colonies in Chembur and the vicinity of India's atomic research reactor, known as Purnima, which is located on the waterfront south of the Trombay Hills.

While local concentrations of factories also exist along the rail line and highway in the western sector Salsette, commutation of white-collar workers by rail to Bombay Island is much more prevalent there. An important reason for the spread of commuter population along this side of Salsette in preference to the Central Railway on the east side is doubtless the good access provided on the Western Railway on which commuter service terminates at Churchgate Station within 10 to 15 minutes walk of the major private business offices and governmental agencies in the Fort and adjacent area towards Marine Drive on Back Bay. The oldest "bedroom suburbs" on Salsette are Bandra and Khar, located west of the rail line 10 to 12 kilometres from the city centre. Subsequent expansion occurred in and around Santa Cruz and Vile Parle up to 15 kilometres, or 40-50 minutes travel time, from Churchgate Station. Juhu Beach to the west has attracted the wealthy class of residents, but as yet the concentration of Bombay's elite remains much less here than that on Malabar and Cumballa Hills. Greater numbers of new factories and larger local employment base exist in the outermost part of the western sector on Salsette, beginning at Andheri about 16 kilometres from the city centre and extending up to Dahisor, 27 kilometres distant.

SOCIO-ECONOMIC STRUCTURE : DELHI

Analysis of the census data for 88 divisions in the Delhi Municipal Corporation, New Delhi Municipality and Delhi Cantonment reveals four factors (Table 6.3). The four factors together explain almost 72 per cent of total variance.

Factor 1, which explains over 24 per cent of the total variance, shows high positive associations between population density and housing concentration with employment in trade and commerce, household industries together with female/male ratio among the Scheduled Castes. This factor also shows strong negative associations with agriculture, livestock breeding and other primary production, construction and services. Factor 1 can be considered to be the measure of the degree to which Delhi retains the features of traditional urbanism with population highly concentrated in and around the central bazars of the old city. Factor 2, which explains some 17 per cent of the total variance, shows a strong inverse association between literacy of

TABLE 6.3 : **Factor analysis—Delhi 1971 (Rotated factor loadings over + or—.300)**

Variables	Factor 1	Factor 2	Factor 3	Factor 4
Persons per hectare	+.829	—	—	—
Persons per house	+.676	—	—	—
Persons per household	+.783	—	—	—.413
Ratio of females to males	—	—	—	—.870
% of population : scheduled castes	—	—.800	—	—
Female/Male ratio : scheduled castes	+.434	—.485	—	—.410
Literacy : % total males literate	—	+.875	+.343	—
Literacy : % total females literate	—	+.885	+.366	—
Total males : % workers	—	—	—	+.908
Total females : % workers	—.399	—	+.773	—
Total workers : % in farming and farm labour	—.410	—	—	—
Total workers : % in livestock, forestry, fishing, and other primary production	—.723	—	—	—
Total workers : % in household industries	+.572	—.441	—.309	—
Total workers : % in manufacturing industries	—	—	—.801	—
Total workers : % in construction	—.591	—.483	—.373	—
Total workers : % in trade and commerce	+.772	—	—	—
Total workers : % in transport and communication	—	—	—.556	—
Total workers : % in services	—.413	—	+.833	—
% of variance explained	24.3	17.4	16.6	13.4

both males and females and Scheduled Caste population. Employment of workers in household industries and construction are also negatively related to literacy in this factor. Although Factor 2 in Delhi differs in important respects from Factor 1 in Bombay, a positive score can be taken as a surrogate of high social status. Factor 3, which explains some 16 per cent of the total variance, also gives a positive association of literacy for both males and females, although the loadings are less marked than in Factor 2. Factor 3 brings out a strong association of workers employed in services and working females. Factor 4, which shows the concentration of male workers is inversely related to female/male ratios.

Geographical distributions of the scores for the four factors by census division provide the basis for understanding of the spatial structure of Delhi. Factor 1, traditional urbanism, shows scores exceeding +1.0, not surprisingly, inside the formerly walled city of Old Delhi and between 1 and 2 kilometres west in Sadar Bazar and Sabzi Mandi, which are both long-settled areas "outside the walls". Other areas more recently established, such as Karol Bagh-Rajendra Nagar and Patel Nagar are removed 2 to 3 kilometres west of the centre of Old Delhi, while one outlying area above +1.0 exists on the northwest on the Grand Trunk Road, 3 to 4 kilometres distant. Generally, the areas with positive scores in Factor 1 occur within 7 kilometres of the city centre on the west and east of the Yamuna in Shahadara within a radius of 9 kilometres. Factor 2, signifying high social status, usually shows scores exceeding +1.0 in areas outside the urban core of metropolitan Delhi. Only one census division with such high status is found adjacent to Chandni Chauk in Old Delhi. The other areas with high social status are removed some 2 to 3 kilometres west in Karol Bagh-Rajendra Nagar, as well as in other divisions 3 to 10 kilometres or more distant in the west and northwest sectors of metropolitan Delhi. This factor does not assume as much prominence in the south sector of the metropolitan area except in one part of New Delhi, some 10 kilometres from the centre of the city. There are positive scores in Factor 2, however, spread over large parts of New Delhi, and the southern, western, northern and eastern parts of the Delhi Municipal Corporation in sectoral fashion.

Factor 3, which depicts the Delhi of military and civil

administration services, as well as the vast range of private professional and personal services, shows that the census divisions with scores above +1.0 occur in close geographical association with Delhi Cantonment and New Delhi Municipality. Except for the Civil Lines, north of Old Delhi, which retain certain governmental functions, the preponderance of census divisions with positive factor scores and found in the southern half of the metropolitan area accentuating the dual geographical structure of Greater Delhi. Male worker concentration, identified by Factor 4, tends also to be most intense in Delhi Cantonment and New Delhi Municipality, where scores commonly exceed +1.0. But equally high male worker concentrations exist in Old Delhi in the neighbourhood of the major railway and bus terminals, the Delhi University Campus, and other localities on the west and east of the urban core in certain factory areas and labour colonies.

SPATIAL STRUCTURE : DELHI

The spatial structure of Greater Delhi is delineated, as in Bombay, by combining Factors 1 and 2. The two factor scores are not inversely related as frequently as in Greater Bombay. Hence, the areas of high social status are not as sharply differentiated from areas of low social rank. Indeed, there is significant overlapping of high social status (Factor 2+) with traditional urbanism (Factor 1+). Thus, areas of positive scores in Factors 1 and 2 are found in the highly commercialized and densely populated areas with high levels of literacy, i.e., the high-status areas of Old Delhi and its vicinity. The areas of positive scores in Factor 1 and negative scores in Factor 2 (1+ and 2—) occur in close geographical association with the innermost area of high social status near the core of Old Delhi outlined by a heavy dotted line (Fig. 6.7).

The other combinations of factors, bring out areas with a low degree of traditional urbanism with high social status (1— and 2+) or areas with negative scores in both factors (2— and 1—), signifying both low traditional status and low social rank. A largely sectoral pattern of the latter two factor combinations forms a broad belt outside the above-mentioned core of Inner Delhi. The largest upper-class sector (1— and 2+) begins in New Delhi 2 kilometres south of Chandni Chauk at Connaught Circle

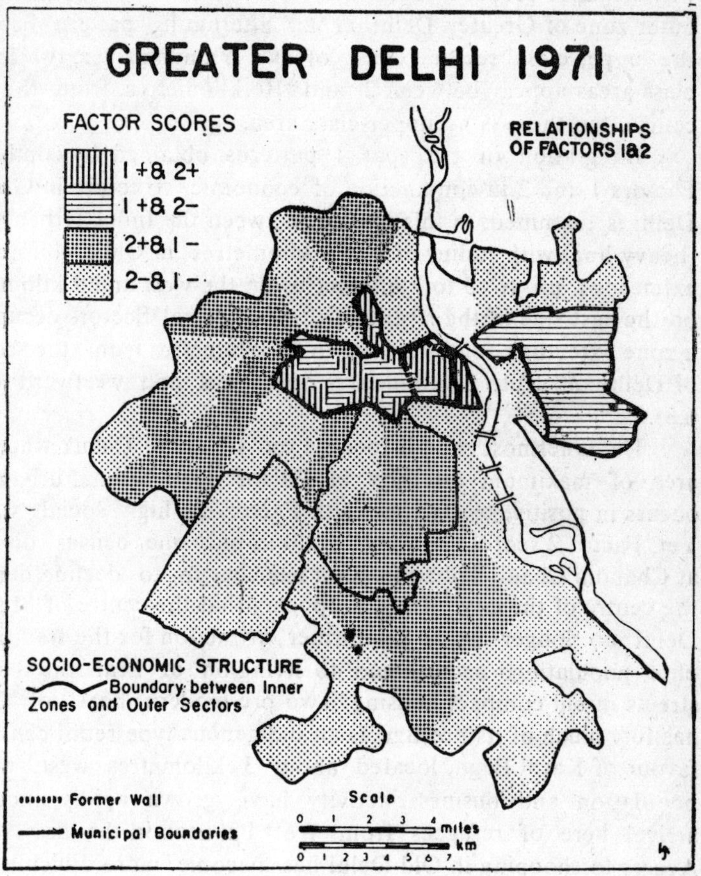

FIG. 6.7 : *Greater Delhi 1971—Relationships of Factors 1 & 2*

and extends southward into the area administered by Delhi Municipal Corporation as far as the southern boundary near the Qutb Minar, 18 kilometres from Chandni Chauk. Two other upper-class sectors are situated in west Delhi between 7 and 12 or 15 kilometres from the centre of Old Delhi. Several census divisions upper-class population are found adjacent to the compact urban area as close as 3 kilometres from Chandni Chauk on the west, while another well-defined sector exists in the north between 5 and 9 kilometres distant. Five main lower-class sectors with negative scores in both factors (2— and 1—) occur in the outer zone of Greater Delhi in an alternating pattern between the upper-class sectors. East of the Yamuna river, two lower-class areas appear between 3 and 10 kilometres from the city centre, but there is no upper-class area.

Integration of the spatial patterns obtained by combining Factors 1 and 2 in conjunction of economic structure in Greater Delhi is examined, in this section, between the Inner Urban Core (heavy line) with radius of 1 to 2 kilometres in Old Delhi and its extensions up to 5 to 7 kilometres on the west and 9 kilometres on the east side of the river and the Peripheral Sectors occupying a zone extending southward up to 20 kilometres from the centre of Delhi and 10 to 16 kilometres northward and westward (Fig. 6.8).

The innermost zone is focused on Chandni Chauk where an area of maximum intensity in Factor 1, traditional urbanism, occurs in positive association with Factor 2, high social status. Yet, Factor 2 scores are above +1.0 in only one census division at Chandni Chauk and population has begun to decline here in the centre of the city. Thus, it appears that the centre of Mughal Delhi no longer holds its former attraction for the traditional elite population, accustomed to living on or near the bazaar streets in old compact housing. Two pre-modern heart of Delhi has lost much of its prestige as an indigenous-type retail centre in favour of Karol Bagh, located about 3 kilometres west, where population and business activity have grown rapidly since the arrival here of refugees from West Pakistan in the late 1940's. Access to shopping in Old Delhi has become more difficult than in Karol Bagh, which is adjacent to new middle-class housing colonies, such as Rajendra Nagar and Patel Nagar, and relatively close to the upper-class population living in West Delhi. Similar growth is observed from Sabzi Mandi northwest along the Grand Trunk Road.

Delhi's Inner Zone contains the principal rail and road transport terminals and its main industrial concentration, giving rise to ring-like patterns of population of low social rank and male worker concentrations. The only large area where relatively unrestricted expansion of housing has developed for lower-class population is found east of the Yamuna river, mainly on the south side of the transportation corridor formed by the main

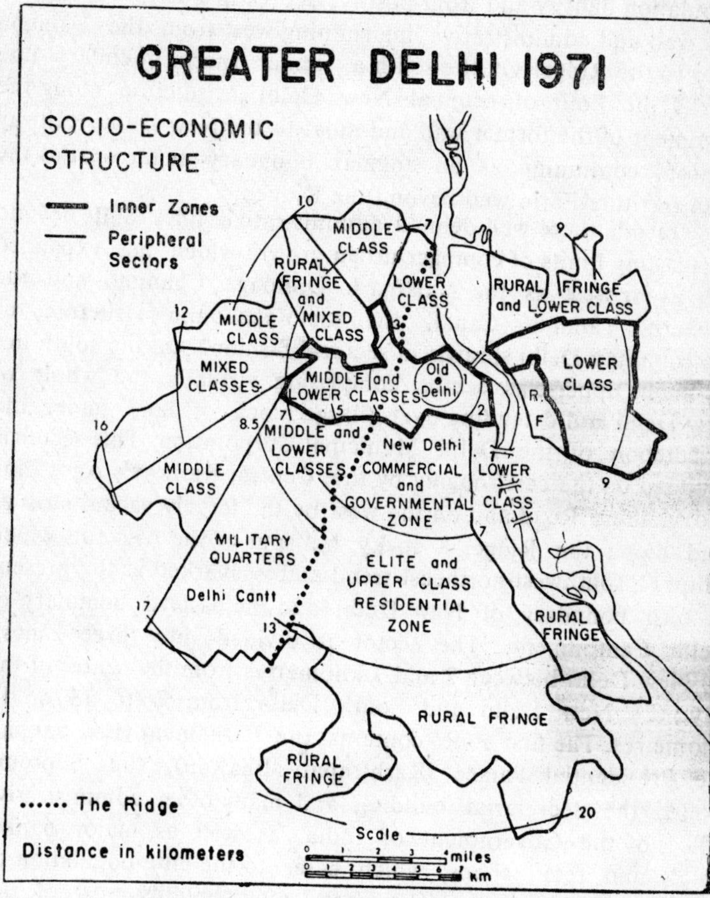

FIG. 6.8 : *Greater Delhi 1971—Socio-economic structure*

Northern Railway line and Grand Trunk Road leading eastward into Uttar Pradesh. West of the river, the spread of a traditional urban landscape has been prevented by restrictions imposed on "natural growth" on the north side by the original 19th century Civil Lines and later by Delhi University Campus and by the establishment of the separate New Delhi Municipal jurisdiction on the south side. The effect of the former boundary of the Delhi Civil Lines can be discerned in terms of the sharp drop of population density and from positive scores in Factor 1 along the city wall and administrative line running west from the Yamuna river to the Ridge. On the south a similar sharp boundary line marks the limit of original New Delhi jurisdiction along the alignment of the former wall and moat to the New Delhi Railway Station, continuing as an angular boundary as far west as the Pusa Institute Farm well beyond the Ridge.

Outside these well-defined administrative lines and beyond the existing limits of compact urban growth which has expanded only as far west as the Najafgarh Drainage Channel, and the Western Yamuna Canal, the socio-economic structure of Metropolitan Delhi sprawls in sectoral fashion. On the south lies the major upper-class sector, comprising virtually the whole of New Delhi and extending over a large block of land under the jurisdiction of the Delhi Municipal Corporation. This sector is bounded on the east roughly by the Central Railway's Agra Line and Mathura Road and on the west by the largely vacant strip of land along the Ridge, a rocky hill spur of the Aravallis which follows a southwest-northeast trend and is marked as the present western boundary of New Delhi and the eastern boundary of Delhi Cantonment. The sector is divided into three zones: (1) New Delhi between 2 and 7 kilometres from the centre of the city; (2) New Delhi and South Delhi from 7 to 13 or 14 kilometres. The first zone contains the Parliament (Lok Sabha), the Presidential Palace (Rashtrapati Bhawan), the Supreme Court, the Secretariat building and many other administrative offices of the Government of India, as well as major banks, hotels and retail shops, which cater to the elite population of the national capital area. The second zone contains most of the luxury and upper middle-class residential colonies, as well as many new hotels and second-level shopping centres, catering to upper-class Indians and foreign tourists. The residents of this sector are able to afford to use private automobiles to an extent seldom observed elsewhere in India with the result that a typical feature of most dwellings in the luxury colonies is a built-in garage. There is a tendency for commercial encroachment to occur in residential structures along the major arteries and local traffic congestion at the rush hours causes one to think of this sector as an Indian counterpart of our American suburbs. Indeed, a major new shopping centre is already under construction beside the Greater Kailash Colony at the extreme southeast angle of this sector. The third zone is the rural fringe, including the outlying villages of Mehrauli and Tughlakabad and other agricultural areas, now merged into the Delhi Municipal Corporation. Upper-class residential expansion is occurring in this zone along Aurobindo Marg up to the low stoney hills where the Qutb Minar stands, overlooking the ruins of pre-Mughal Delhi.

The New Delhi-South Delhi sector is flanked by two lower-class sectors. On the east is a long strip of land under jurisdiction of Delhi Municipal Corporation, running 14 to 15 kilometres along the Yamuna river from the former city wall in the north to the rural fringe in the south. This sector contains some low quality housing and governmental or private institutional offices in the north, but it has been intruded by some high-quality housing colonies in the mid-portion. The lower-class sector on the west is divided into two zones. The first, lying between 3 and 8 to 10 kilometres from the city centre, is characterized by spotty distribution of good and poor quality housing, urbanized villages, the Pusa Institute Farm, and open land occupied by jungle and stone quarries. The second zone beyond some 8 kilometres and up to 17 from the city centre comprises the Delhi Cantonment with housing and associated facilities for military forces in the northern part and the Delhi International Airport in the southern part.

Three more upper-class sectors are observed in the western and northern quadrants of Greater Delhi in an alternating arrangement with sectors of mixed or low social rank. Beginning in the southwest is a high-status sector, containing Rajouri Gardens and several other middle-class residential colonies, located between the road to Najafgarh and the rail lines bounding Delhi Cantonment. On the other side this sector borders a largely-empty sector with a mixed population of predominantly low rank. Next comes a high-status sector between Rohtak Road and the Western Yamuna Canal; followed by a largely agricultural lower-class sector, extending up to the Grand Trunk Road. On the north side of the Grand Trunk Road up to the Najafgarh Drain is small high-status sector, containing Model Town and other middle-class housing colonies. Lastly, there is an anomalous lower-class sector, comprising Delhi University and the old Civil Lines, where evidently the educated elite, engaged with higher education and administration and the courts of Delhi State and outnumbered by people of low social rank, who spill over from the adjacent congested Inner Zone.

The small lower-class sectors appearing on the east side of the Yamuna river are located either to the north or the south of the main line of travel, and are occupied mostly by labourers and urbanized villagers. Local public transportation by buses and railways is little developed in the Delhi metropolitan area, as compared to Bombay. Costs of travel by motor rickshaw or taxi is prohibitive for the vast majority of the inhabitants. The result is heavy dependence on bicycles and walking, which puts severe limitations on commuting distances for the poor of Delhi. Urban expansion in the western, northern and eastern quadrants is confined to a radius of 10 kilometres and most people must live within 3 to 5 kilometres, which is one hour's walk of their work places.

CONCLUSIONS

Factor analysis of the population in Bombay and Delhi shows that the Bombay is more clearly stratified in upper and lower classes. In essence, there are two Bombays—one for the elite and upper-class with education and high achieved economic status; the other for the mill workers and labourers on the docks and railways with low social rank and little education. In Delhi there is a three-way separation of the upper-class population. The traditional urban population of merchants, bankers and educated people with high social status in Old Delhi is contrasted, on the one hand, with the new upper and middle class merchants and other entrepreneurs of largely Punjabi refugee origin in Karol Bagh and West Delhi and, on the other hand, with the more diverse elite of New Delhi and South Delhi associated with the functions of India's national government. Population of lower class is confined either to congested areas of the Delhi urban core and its immediate vicinity or in lesser numbers to the peripheral sectors and rural fringes.

Geographical analysis of the socio-economic structure of Bombay and Delhi demonstrates that there are inner zones of old upper-class concentration and outer sectors of recent upper-class expansion. The inner zones in both urban areas have predominantly concentric spatial patterns with the lower classes living farther from the centre than the upper classes. However, the tendency towards centralization seems to have passed its most intense phase. Population decline in the last census decade presages the loss of former dominance of the city centre as a preferred place of residence. The naturally confined insular site of Bombay and the administratively restricted site of Delhi contribute towards the maintenance of high population density and compact urban development. However, the outer zones where most of the recent population growth has been accommodated are developing in sectoral patterns because of the influence of transport corridors, accessibility for daily travel to work and local site factors. Thus, the urban cores of both cities retain features which reflect traditional preference of the elite for central residential location, while current growth towards the peripheries parallels the geographical pattern of so-called Western urbanization.

SEVEN

JAY A. WEINSTEIN

MADRAS
An Analysis of Urban Ecological Structure in India*

THE DEVELOPMENT OF THE ECOLOGICAL APPROACH

EUROPEAN social theory of the nineteenth century focussed upon the city as the most likely and most dynamic ground for the working out of social-relationships of the industrial age. Major insights of classical urban sociology include the identification of institution-related "settings" (e.g., market place, fortress, religious centre) as major determinants of urban life and the attribution of increasing complexity to the social and, by implication, the ecological structure of a modern metropolis. But, if one were to characterize the development of the field of urban ecology itself, from the original contributions of Park, Burgess, and McKenzie (1925), (Park 1916) and their colleagues of the Chicago School, through the work of the social area analysts, to studies done in the past few years, he or she would find a clear and steady trend *away* from the making of a priori assumptions about the nature of the sociogeographic organization of the city: the trend has been toward the discovery of the real or "natural" subunits of which the city is composed, independent of theoretical disposition or bias incorporated, for example, into the research instruments (for a review of these developments see Schwirian) (1974).

The most recent development of "assumptionless" technology

*Reprinted from *Sage Research Papers in the Social Sciences*, Series Number 90-115, Vol. 2, 1974, by permission of the author and editor.

in urban research arose partly out of the incomparability of census procedures in non-Western societies and partly for less obvious scientific reasons. Factorial ecology in the broad sense, as the approach is known, employs factor analysis on a set of variables which may turn out to be quite different from those used in social area analysis. What variables are, in fact, used depends upon a combination of two research priorities : (1) the kinds of data available, and (2) the research interests of the investigator. Because these things vary from study to study there is a disappointingly high degree of incomparability between studies employing this technique. Most ecologists have spoken to this point and it is of central concern in what must count as the outstanding review of the field to date (Schwirian 1971).

There seems to be a nearly inextricable dilemma in this regard (crucial practical problems aside). Factorial ecology comes closest to producing an unbiased structural description. If one were to legislate the variables to be employed one might be imposing a debilitating preconception. Yet, without such standardization, sophisticated comparison becomes more a matter of art than of science. In any case, it is now general practice to employ "open" factor-analytic techniques, thereby finding (or inventing) the compound variables which explain the greatest amount of variance among the data which have been accumulated for a particular city (and, wisely, leaving the naming of the factors to the reader). Once the factor scores have been produced they are associated with geographic variables, or referred back to the map, in order to display spatial-ecological structure.

INSTITUTION-RELATED SETTINGS AND COLLECTIVE REPRESENTATION

Madras is essentially a political city. It is the fourth largest city of India, the State Capital of Tamil Nadu, and the former administrative capital of Madras Presidency. By 1971, Madras had a population of 2.5 million and an area of 49.7 sq miles (In 1961 Madras was divided into 100 census divisions and in 1981 into 120). Coimbatore to the west and Madurai to the southwest are the major industrial and religious cities respectively of the Tamil-speaking population of India.

Nevertheless, Madras has significance for Indian religion and industry. There is, for example, a substantial industrial sector which includes major textile mills and the national centre of railway coach construction. In addition, Madras is the site of the world centre of the Theosophical Society and of a pilgrimage centre for Indian Christians. Of special importance to Christians is Mylapore, now a section of the city, in which is located St. Thom Cathedral, a Roman Catholic Church, the first in Madras. This is the Church of Saint Thomas the Apostle, patron saint of India, who is said to have settled in Mylapore in the first century A.D.

Despite the fact of early Christian settlement in the area and indications that it already had been well established as a pilgrimage centre by the time of St. Thomas' journey, Madras, as such, is an invention of British colonial policy and the activities of the British East India Company.

Through dealings with the French, among others, land was acquired at the mouth of the Cocum river where Fort St. George was established in 1642. Behind the fort and to the north was the bazaar town of Georgetown (Sennaipatinaam). To the south was Mylapore. With time, the fortress settlement joined, through expansion of all three populations, with the bazaar and pilgrimage centres to become (with a few nearby settlements) Madras.

The growth of Madras, and especially the conjunction of the Bazaar, Fortress, and Temple, has a parallel in the theoretical discussion and propositional inventory which follow. Indeed, it is the relationship between these theoretical considerations and the historical data which suggest formulation of the specific research hypotheses tested here. For, by the three centre model which we propose, an adequate test of expectations about Madras' ecological structure would require, at least, a consideration of the influence of each of these settings upon the geograpic distribution of social characteristics.

If there is any "definitive" ecological structure for Madras and if this ecological structure is "setting associated", then, following the hypothetical directions suggested. Patterns of land use, land value, and geographic distribution of social groups are matters of tri-dimensionality or, projected to a plane, triangulation.

Significant direction in developing this notion is to be found in Bell (1969) and Berry and Rees (1969). Specifically, social area analysis (of which Bell speaks) considers the spatial relationships between social areas and the indices taken three at a time. Berry and Rees discuss and demonstrate, to some extent isomorphism (and its absence) between "trend surfaces" for various factors. In the case of our study of the ecological structure of Madras, these matters were considered in the light of the three-setting principle. Not only are there three separate clusters of social characteristics (operationalized as social area indices or factors) but each may be distributed geographically in relation to the principal setting which is a space-bound representation of the institutional structure with which the cluster is associated.

Pre-Weberian urban theory was guided by two principal postulates :

1. The existence of industrial, and class, society implies the diminution of the importance and, ultimately, the (population) size of the country side.
2. The increase in size and importance of the urban sector implies a simplification of structural (class) relations.

Taken together, these postulates underlie an "urban determinism" which was characteristic of most nineteenth century social theory. The problem for comparative urban studies today lies in separating that which is urban from that which is industrial, Western, or "modern". With Weber's work (1958) (particularly his commentary on the Bazaar-Fortress polarity) a foundation was laid for a better understanding of urbanization in non-Western contexts. The most likely source of these postulates, and the locus of Weber's revision, lies in the assumption that economic relations will overwhelm all other in that collectivities whose basis of association is not adjusted to the rationale of the economy cannot possibly survive.

Weber's addition of three political conditions (of his five-condition set) basic to the existence of a city and, especially, his elaboration of the role of the fortress in the history and structure of the city, precisely provides the palliative for an economic determinist urban theory. Yet, in the work of Park and Burgess the "central" business district—the Bazaar—has been treated as a unique "centre" for ecological models. If Weber's commentary on the fortress represents a true advance in theory, then the choice of the central business district as unique is a regression. To include Weber's insight, ecologists would focus on the influence of the fortress, as well as the market, upon the distribution of social characteristics in urban space, especially in accidental cities (like Chicago).

The most probable source of this step backward is that neither Weber nor his contemporaries seemed particularly concerned with relating the historical and social-structural significance of the market (or fortress) to the inner dynamics of urban social geography, although Berry and Rees cite a "mid-nineteenth century classic" on urban sociology in which the author, Johan Georg Kohl,

> . . . devoted an entire chapter to the internal structure of cities. Urban structure, he said, might be viewed as a series of layers. . . . These layers . . . translate into "arches" of homogeneous social structure so that economic and social importance (*sic.*) of the population declines as you move out from the centre of the city (Berry and Rees 1969).

The real task of establishing a formal link between classic insight and ecological structure of which this quotation is a harbinger (Kohl's model characterizes a situation nearly the exact-reverse of the Park and Burgess concentric zone model) fell to the Chicago School.

In an indirect way another important part of a foundation theory of urban structure comes from Durkheim, although he, too, failed to make explicit spatial-ecological application. Principally, there are two insights which constitute Durkheim's contribution.

The first of these entails the argument of *The Division of Labour in Society*. Briefly, to a significant extent it is established that urban social structure (perhaps everywhere, perhaps particularly in the Third World) is complex compared to the Marxist expectation for an advanced capitalist and highly polarized structure. Part of the reason for this entails the high division of labour, the high degree of specialization, and the relative predominance of organic solidarity associated with "modern" cities. Thus, if we can assume a clear association between social structure and ecological structure, it would follow from Durkheim that the ecological structure exhibits a complexity consonant with the complexity of the social structure.

The assumption of such a clear association is supported by the second and more significant contribution of Durkheim, the concept of "collective representation". To quote :

> Collective representations are the result of immense cooperation, which stretches not only into space but into time as well ; to make them, a multitude of minds have associated and combined their ideas and sentiments . . . (Durkheim 1954).

The connection may be apparent. The ecological structure of a city is a collective representation.

With this link, we would suggest, the direction for the use of ecological theory in comparative research becomes clearer, for if ecological structure is a collective representation, it is nearly unique among sociological phenomena of like magnitude. It is undeniably concrete—unlike institutional structure—in that it is bound in space and time.

Propositions and hypotheses

In assuming that the central business district, the bazaar, be treated as the centre of ecological structure, the Chicago School's concentric zone model employs a set of economically determinist assumptions. In detail, this model assumes that the bazaar exerts a "pull" on the settlement pattern of individual households so that where a household locates relative to the bazaar is in part a function of the social characteristics of that household. Since the bazaar is distinct, distinct regions form at varying distances from it in terms of desirability. Distance from the bazaar is not in this model a strictly linear concept. Instead, there is adjacent to the bazaar a highly desirable neighbourhood (the Gold Coast), then the next (or, perhaps, equally) most desirable area is that which is farthest away ; finally, in successive zones from the outskirts to the centre, desirability decreases up to the transitional zone. Since this is the most widely accepted account of concentric ring formation, we accept it as a general proposition regardless of which setting (or imaginary point) is chosen as centre. Thus, where there are five zones around the bazaar (or any centre), their order of desirability from the centre outward) is : 1, 5, 4, 3, 2.

TYPES OF DATA

Three types of data in the analysis of Madras' ecological structure have been employed : sociogeographic, strict demographic, and social structural.

Sociogeographic variables

Sociogeographic variables characterize an areal sub-unit, that is division by referring to its space relations to a setting.

1. By Bazaar Setting it is meant that area which is generally counted as the main business district.
2. By Fortress Setting it is meant the foundation fort of a city.
3. By Temple Setting it is meant to specify a religious building which satisfies as many of the following criteria as possible : (a) the first such building in the history of the city ; (b) the main building of the major religious group in the city ; (c) the most important pilgrimage centre of the city.
4. In measuring distance from a setting to an area the convention of taking the straight line distance (except where angular measure is called for) to the geographic centre of a census tract is followed.

Strict demographic variables

Strict demographic variables are what one commonly thought of as census variables and are the kinds of variables included (possibly among other kinds) in all national census reports. A strict demographic variable has two defining features :

1. It is presumed to be based upon a complete enumeration.
2. It characterizes an area (in the case of ecological studies, a census division) by its population of individual components : individual human beings or households.

Social structural variables

Anticipating the possibility that even the most sophisticated techniques of factor analysis would not provide indices of the sort required by the conventions of factorial ecology or by the research hypotheses, supplementary information was gathered on 2,000 households in Madras by way of an interview survey. Thus, there have been several social structural variables with which to compare the strict demographic data. The sampling procedure was based on the assumption that generalizations could be made from the households to the block, from the block to the division, from the division to the social area and, by way of theory, from the social area to the city.

THE ECOLOGICAL STRUCTURE OF MADRAS—SOCIOGEOGRAPHIC

The purpose of this section is to outline certain of the characteristics of Madras referred to in the discussion of types of variables as "sociogeographic". If the conception of the process of Madras' growth as the movement of households into the immediate neighbourhood of the bazaar, fortress, and temple can be retained and, further, if the ongoing dynamics of this dimension of city life in terms of the constant movement of households within (and out of) this region can be considered the ecological structure can be conceived of as a frozen moment in this process. The ecological discipline assumes that the space over which the process occurs is regional (anisotropic). In particular, the assumption goes, where a household is located at a given moment depends upon a combination of three sets of characteristics : (1) those of the household ; (2) those of other households in the same location ; and (3) those of the location, independent of the households which it contains. Although the characteristics in the third set are independent of household characteristics, they are related to other aspects of the city which are identifiable. Some of these, such as proximity to a river or railway tracks, height above sea level, or the number of parks, are important but are not treated in detail in this paper. Others, such as proximity to the bazaar, are the subject of this section.

In this sense, then, the next few pages as the framework upon which the structure of inquiry has been built can be treated just as, following the ecological bias, its subject is the framework upon which the ecological structure of Madras has evolved.

The bazaar and its concentric zone system

At the eastern extremity of the city, about two miles south of the northern boundary, is Madras harbour. Separating the harbour from the city proper is First Line Beach Road. To the south, this road becomes South Beach Road. A white sand beach runs the length of the city—the second longest beach in the world (after Cox's Bazaar). First Line Beach Road is the easternmost boundary of Parry's Corner, Old Georgetown, and the Bazaar of Madras. Since this is an extensive region, we have calculated distances from its approximate centre, the intersection of N.S.C. Bose Road and Armenian Street, the location of several banking houses and other financial centres of South India. From this centre, we have constructed a concentric zone system of five circles, the first with a radius of ½ mile and successive ones with radii incremented by ½ mile. Map 2 locates this system and indicates, by number, the divisions whose centres fall within the various zones.

According to ecological theory, these zones ought to form distinct regions of varying desirability along some socio-economic dimension. In order to approximate a test of this proposition, we asked the 2,000 heads of household sampled to name three choices of neighbourhoods in which they would like to live (including their present neighbourhood). By classifying these preferences by division we were able to assign to each a desirability score. We then classified the divisions into five ranked categories and compared these with the hypothetical classification which follows from the bazaar-centred concentric zone system. Results are discussed below.

The fortress-centred concentric zone system

One and three-fourths miles to the south and east of the bazaar, along South Beach Road just as it turns toward the sea, is Fort St. George. The Fort is legendary in the popular culture of Madras largely because no shot has ever been fired from it for the purpose of combat. Nonetheless, it is an extensive series of connected buildings, comparable to Fort William in Calcutta, set on a rise so as to have a commanding view of the sea. It is a major tourist attraction of the area and today it serves as the seat of the Tamil Nadu State Government.

By constructing a set of concentric circles at ½ mile increments from the Fort, we established the second, fortress centered-concentric zone system. The relative location of the division centres was compared with the desirability ratings in a manner identical to that employed with respect to the bazaar.

The temple-centred concentric zone system

Approximately three and one-half miles nearly due south of Fort St. George is the section of Madras which was the old town of Mylapore. Here there are located at least two unambiguously principal (and several likely) religious centres of Madras. The first of these is St. Thom Cathedral and the second is Kapalaswara Temple. In addition, one mile to the southwest of Kapalaswara, in Adyar, is the world centre of the Theosophical Society and one mile to the northwest of St. Thom is Amil Mahal, a Muslim shrine and Mosque. For these reasons we have chosen to consider this area as the focal point of Madras "Oriental-City" tendencies. That is, for this area at least, we would suggest,

> the principle of residence never superseded the significance of kinship ties as a basis for social organization. Consequently, the Oriental city represented a more or less loose collection of kinship and tribal groups who monopolized particular skills and trades for themselves excluding all outgroup members (Murvar 1966).

A point halfway between St. Thom and Kapalaswara for the temple setting has been chosen. Around this a concentric zone system has again been constructed. Once again, this system of zones was compared with the desirability ratings. To the extent that the above quoted passage applies to Madras and to the extent that this desirability rating is most clearly associated with a prestige dimension, it may be suggested that the agreement between hypothetical rank and actual rank ought to be closest for this concentric zone system.

The centroid-centred concentric zone system

A fourth and final set of concentric circles is about an imaginary point which is the intersection of the medians of the triangle connecting the bazaar, the fortress, and the temple. Interestingly, this point lies on The Island Grounds, an island in the Cooum river which serves, among other things, as the site of a recreation area and the Madras City Jail. The Grounds are encircled by water and, beyond, by the central city ; so its geographical character points to its theoretical significance to the ecologist.

Sectors of equal distance from pairs of settings

By constructing the perpendicular bisectors of the sides of the triangle joining the bazaar, fortress and temple it was possible to establish a theory-related sector system. Because of the proximity of the bazaar and the fortress these three lines roughly divide the city into three parts with a north-south orientation. In constructing the sectors all divisions whose centres lie within 15° of these lines have been classed together. Three principal sectors resulted from this construction which, again in a rough way, cioncide with :

(1) the area lying between the northern city limits and Poonamalle High Road ; (2) the area lying between Poonamalle High Road and Mount Road ; and (3) the area lying between Mount Road and the southern city limits. Since these two roads are (along with South Beach Road and its northern extension) undoubtedly the principal arteries of Madras, there is reason to believe that there is a foundation for treating certain aspects of the ecological structure as sectoral. That none of the census variables or household survey factors is related to this construction is partly understandable in light of the fact that for people who do not own automobiles (the overwhelming majority of Indians), living far from the central city but close to a road is no advantage in terms of access, as it is in North America where the model was developed. The divisions lying in the respective proximal sectors were compared with those ranking high in desirability.

The geographic distribution of "Desirable" divisions

From the comparisons between hypothetical and actual class

membership, in which there is approximately 10 per cent exact agreement and 30 per cent maximum disagreement, one may conclude that the "desirability" of divisions does not correspond to zonal or sectoral patterns (although the suggestion that the temple-centered system ought to produce the closest agreement is found to be warranted). Thus, either the question on choice of localities and/or the related indexing are not valid indicators of desirability along any dimension or the theory, or the interpretation of it is suspect.

There are several grounds upon which the operational definition may be questioned. Primarily, the indicator of desirability undoubtedly compounds a number of factors which are out of keeping with the spirit of the theory upon which it is intended to shed light. It is likely that the interview subjects have based their preferences upon settled neighbourhoods. There is, further no theoretical reason for believing that there exists a meaningful correlation between the desirability of a location and the desirability of the people who settle in it (but there is a theoretical reason for believing, for example- that there exists such a correlation between the desirability of a location and the social rank of the people who settle in it). Furthermore, if it can be assumed that people desire to live either near their present neighbourhood (and there is reason to assume that this is so) or near those whose social characteristics are much like their own, then the desirability index as presented ought to correlate highly with the characteristics, or the location of the latter. And it does.

In light of this, the only really strong bit of evidence against the theory that one may wish to do away with the "gold coast" assumption is groundless; one may thus treat the functional relationship between desirability and location of zones as linear. That is, by the results, Zone I is somewhat less desirable than the others. This, of course, may have to do with the lack of precision inherent in the use of the census division as a unit, that is, the "gold coast" may have been missed because, by the Park and Burgess model, it would tend to be a small region easily missed as it constitutes only a portion of some of the inner divisions. In any case, it is best to conclude that the test of desirability is interesting and somewhat novel and it may have further use in understanding other aspects of the inner dynamics of the ecological structure of Madras and other cities, but it has little to do with any concentric zone or similar system of which one could conceive.

THE ECOLOGICAL STRUCTURE OF MADRAS—STRICT DEMOGRAPHIC

In this section the results of the analysis of 1961 and 1971 census data for Madras,—data, which have been called "strict demographic" have been presented and discussed. This analysis relies upon factor analytic techniques.

Three kinds of factor analysis have been performed each based upon different theoretical assumptions about the relationship between the matrix to be analyzed (correlation coefficients in this case) and the general factor(s) which "underlie(s)" its structure. The results obtained on the first four factors are presented in this section.

Factor I : social rank

The hypothesis suggests that households highest in socio-economic rank will be found in the concentric zone closest to the bazaar, households second in rank will be found in the concentric zone most distant from the bazaar, and households of progressively decreasing rank will be found in zones progressively closer to the bazaar (from the most distant). In terms of the operational definitions and with the bazaar-centered zone system summarized, this proposition becomes the following research hypothesis :

> *Research Hypothesis I.* Divisions of Madras which rank highest in score on Factor I of a factor analysis of selected census variables lie in Zone I of the bazaar-centered concentric zone system. For other rank classifications and zones, the following relationships hold : divisions ranking second highest lie in Zone V ; divisions ranking third highest lie in Zone IV ; divisions ranking fourth highest lie in Zone III ; and divisions ranking lowest lie in Zone II.

For the first measure of social rank, scores on Factor I have been chosen. Since it explains more than 22 per cent of the variance, it is a relatively potent factor (by chance alone, a factor in this type of analysis would explain about 7%). In this regard, it compares quite favourably with the first factor produced in the 1969 study of the ecological structure of Calcutta (Berry and Rees 1969). It is, however, slightly less potent than the first factor found in employing 1961 census data for Madras. This explained 24.5 per cent of the variance (these percentages refer to the principal axis factors, unrotated). One reason for this difference between 1964 and 1971 is that, because the earlier census included data on slum concentration not repeated for the more recent enumeration, the factor structures are actually different for the two years. This difference aside, the variables with the five highest loadings are the same for both years, though the loadings are more nearly equal for 1971.

TABLE 7.1 : **Factor I—14 selected variables, 1971 census**

Variable	Loading
Principal Axis ; 22.287% Variance Explained	
14.310% Variance Explained	90° Rotation
Per cent literate females	—.755
Per cent literate population	—.750
Workers in manufacturing	.664
Workers in household industries	.597
Per cent houseless	—.485
Centroid : 19.990% variance explained	
Per cent working females	.648
Per cent literate population	.596
Per cent working population	.567
Workers in household industry	—.558
Square Root : Pivot on maximum loading	
Per cent literate population	1.000
Per cent literate females	.901
Per cent in household industry	—.444
Density	—.384

Factor I is a likely social rank index. That is, according to its composition if an area is characterized by many literate females, many literate people generally, few workers in manufacturing or household industries, few members of scheduled castes, and many houseless people it can be characterized as being high in a latent characteristic called social rank. This need not rest solely on a theoretical basis, at least not for this factor, for by choosing variables which clearly indicate social rank in divisions in each of five categories included in the household interview survey, the results are remarkable for the strength of empirical varification they lend to this factor as a measure of social rank.

The results of the test of this hypothesis under consideration may be examined in some detail (See Fig. 7.1).

It is not difficult to judge by a glance at the figure that even by the most generous interpretation, the social areas related to social rank do not coincide with the concentric zone system specified (or with any concentric zone system around the bazaar).

There is a tendency for neighbourhoods of like social rank to cluster together but for their overall organization to display a geometric configuration no more regular than, say, multiple nuclei. Part of this tendency is understandable in terms of the roughness of the census division as a unit, for in examining the data at the block level we find much more homogeneous clustering. This does not account for the fact, which is probably the main fact of the social rank structure of Madras, that within a division, a block which indicates extremely high social rank may just as likely be adjacent to a slum as to another wealthy block. Clearly, there is geographic segregation operating in Madras but the principle upon which it is based appears to be distinct from that apparently operating in Western cities. The literal application of Western models to non-Western cases has not, as yet, provided a basis for exact comparisons, which makes it difficult to draw any but general conclusions about this difference.

Factor II : family structure

The hypothesis also suggests that households "highest" in some social attribute rank are found in the concentric zone closest to the fortress, households second in rank are found in the zone which is most distant from the fortress, and households of progressively decreasing rank are found in zones progressively closer to the fortress (from the most distant).

When factorial ecology has been applied in non-Western societies, Factor II has been assumed to relate to urbanism or familism and, we would argue, these are likely names for it is based upon the structure of the second factor in this study. For

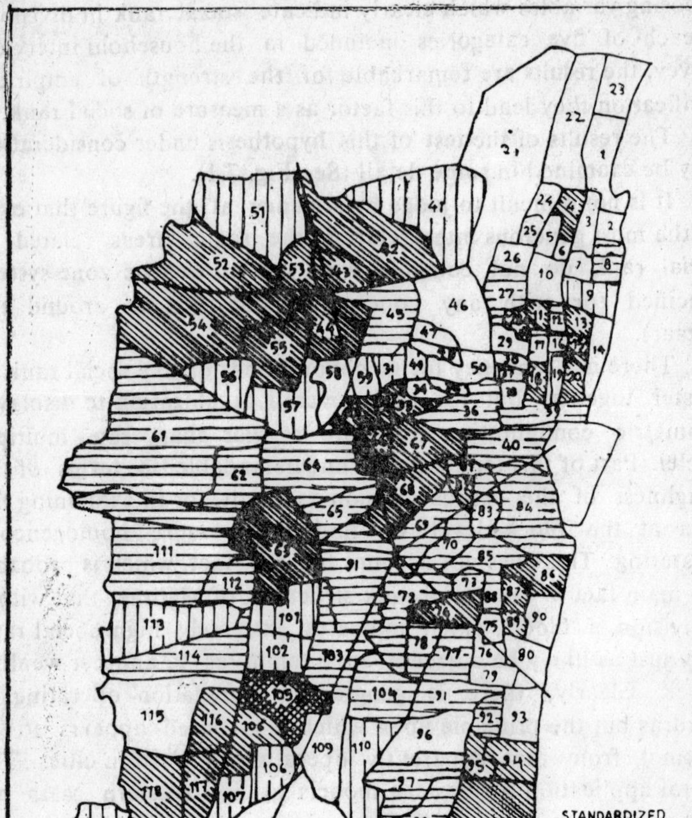

FIG. 7.1 : *Madras City—Social Rank*

this reason, the comparison between the structure of this factor for each of the two census years is an especially interesting one.

In terms of the actual operational definitions and the

determination of the fortress-centered system the research hypothesis related to Factor II becomes :

> *Research Hypothesis 2.* Divisions of Madras which rank highest in score on Factor II of a factor analysis of selected census variables lie in Zone I of the fortress-centered concentric zone system. For other rank classifications, the relationships specified for Research Hypothesis hold in a parallel manner.

For the principal axis analysis, Factor II explains 18.5 per cent of the variance among the data from the 1971 census. Compared to the 15.2 per cent explained by the second factor employing the 1961 data, we have evidence for believing that some progress toward a clear family structure dimension may have been made. This is further substantiated by a more detailed comparison between these two factors which shows differences in the desired direction. For 1971 the variables with the highest loadings are per cent working population, per cent houseless (again occurring as a variable positively correlated with what might be considered welfare variables), per cent working females, and workers in trade, transportation, and commerce.

For the sake of comparison between the two years the list of the outcome of the square root analysis performed for the 1961 and 1971 data has been given. The second factor pivots (by the criterion of maximum loading) on "per cent literate females" whereas for the more recent year the pivot is on "per cent literate population". Since the correlation coefficient between these two variables is .971 and .901 for the two years, respectively, this difference is trivial. There is an interesting observation which might be made on this difference. If one considers the equality of women to be a characteristic of modernity, the fact that the difference between the years of education for males and for females has increased (though slightly) would indicate that Madras has become less modern ! (This is another small bit of evidence in support of the contention that modernity, urbanization, and the passage of time are not, for India at least, necessarily identical things.)

Fig. 7.2 and Table 7.2 summarize the information required for a test of the hypothesis. In a comparison of Fig. 7.2 with a

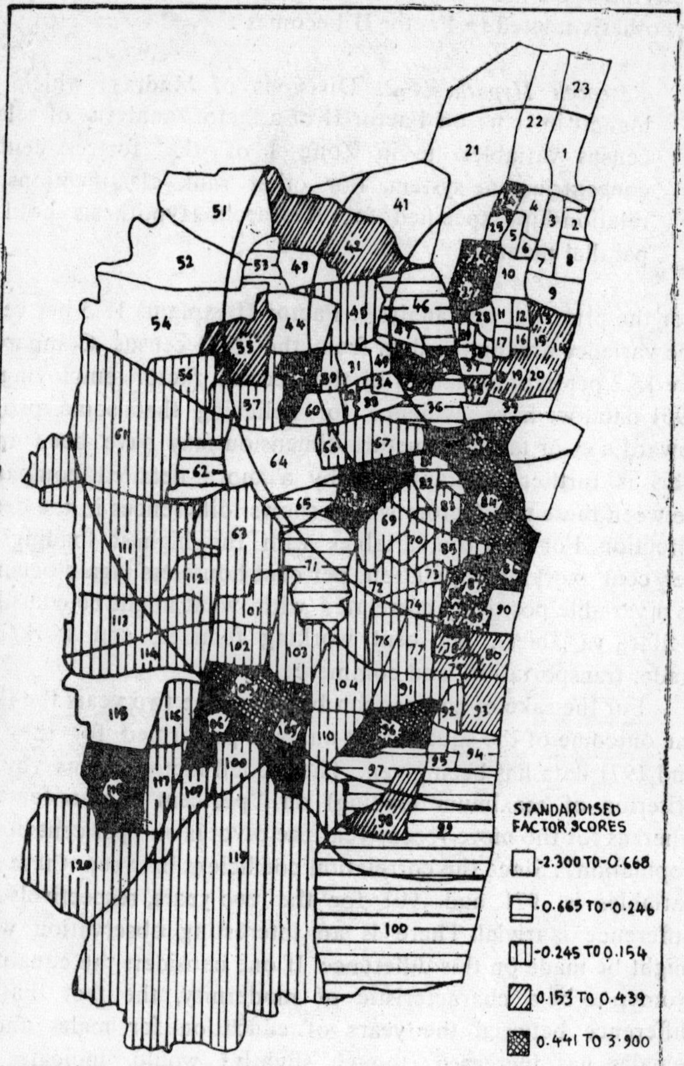

FIG. 7.2 : *Madras City—Family Structure*

map of fortress-centered concentric zone system one may conclude that there is no meaningful relationship between scores on Factor II and the fortress-centered concentric zone system.

TABLE 7.2 : **Factor II—14 selected variables, 1971 census**

Variable	Loading
Principal Axis : 18.507% Variance Explained	
11.528% Variance Explained	90° Rotation
Per cent working population	.786
Per cent houseless	.666
Per cent working females	.663
Workers in trade, transportation, etc.	.663
Centroid : 17.423% Variance Explained	
Per cent working population	—.626
Per cent houseless	—.601
Persons per household	—.469
Per cent literate females	—.437
Square Root : Pivot on "per cent literate population"	
Per cent working population	1.000
Per cent working females	.742
Per cent houseless	.610
Workers in trade, transportation, etc.	.419

In this instance, unlike that of Factor I, the questionable validity of the index leads one to be more reluctant about inferring a weakened theory. There are reasons to believe that family structure displays an ecological structuring different from the concentric zone model hypothesized and, indeed, not at all that of a regular form. To the extent that the interview survey data point in any direction, they support this belief. If this belief is well-warranted, relatively serious implications follow since family structure is the one index among all three (or four) which is held to display a consistent concentric zone formation (Kirschenbaum 1969). This would lead one to strengthen the provisional conclusion that a different principle operates in Madras.

Factor III : ethnicity

The hypothesis also specifies that households highest in rank with respect to ethnicity are found in the concentric zone closest to the temple, those of second highest rank are found in the most distant zone, and households of progressively decreasing rank are found in zones successively closer to the temple. Of the three hypotheses under consideration in this chapter this seems to be the most highly suspect on theoretical grounds. The drift of ecological studies beginning with Firey's (1947) initial work on symbolic ecology suggests no such ranking for ethnicity. This is understandable in light of two aspects in the development of social theory. The first has to do with the reluctance of sociologists to rank ethnic groups as such (as opposed to ranking ethnic groups according to, e.g., mean income), and the second is related to the failure of the early sociologists, to consider ethnicity as a key to the social structure of the modern city. The first of these considerations has provided the basis for the multiple nuclei model which conceives of ethnic neighbourhoods as internally homogeneous but scattered about the city in no particular overall pattern. The second consideration has led to the exclusion of a sociogeographic centre, a "setting" such as the temple which, as in the case of the central business district in relation to social rank, would provide a point of reference for an ordered structuring. The hypothesis under consideration here, to the contrary, specifies both a rank ordering and a centre.

Unfortunately, at least with the use of strict demographic variables, there is no clear way of deciding between the alternatives implied. It does appear that the ethnic component of the ecological structure of Madras displays a multiple nuclei formation. But by the same criteria, so do the social rank and family structure characteristics. With regard to the latter two features, it is possible to conclude that multiple nuclei formation indicates no relationship between demographic and sociogeographic structures. Thus, to be consistent, one ought to do the same for ethnicity ; however, the literal expectations of the theory which underlies ethnic organization are, in fact, less stringent than are those for social rank or family structure. For this reason the hypothesis has been framed in a way parallel to that in which the preceding hypotheses are framed :

> *Research Hypothesis 3.* Divisions of Madras which rank highest in score on Factor III of a factor analysis of 14 selected census variables lie in Zone I of the temple-centered concentric zone system. For other rank classifications and zones, the relationships specified in Hypothesis I hold in a parallel manner.

To restate an a priori consideration in hypothetical form, it is suggested that a more likely order would find divisions which rank second highest in Zone II, those which rank third highest in Zone III, etc. That is, the relationships is more likely to be linear.

Of all the factors considered, Factor III is the least likely to be a valid measure of the concept which it is intended to measure. In so far as it explains 15.3 per cent of the variance (compared to 12.8 per cent explained by Factor III of the analysis of 1961 data) and is thus more potent than Factor II of 1961 data, it is of some technical interest. That it is not an unambiguous measure of ethnicity is clear. The variables with the highest loadings on Factor III are workers in other occupations, density, workers in agriculture and extraction, and persons per household (Table 7.3). Indeed, this may turn out to be the better index for urbanism.

TABLE 7.3 : **Factor III—14 selected variables, 1971 census**

Variable	Loading
Principal Axis : 15.267% Variance Explained	
9.000% Variance Explained	90° Rotation
Workers in other occupations	.816
Density	—.727
Workers in agriculture and extraction	.594
Persons per household	—.384
Centroid : 17.427% Variance Explained	
Workers in other occupations	—.642
Density	.607
Workers in manufacturing	—.537
Workers in trade, transportation, etc.	—.478
Square Root : Pivot on "per cent working population"	
Workers in other occupations	.992
Density	—.454
Workers in agriculture and Extraction	.461

Fig. 7.3 summarizes the information on this factor. Comparisons made between these and the map related to the temple-centered concentric zone system, prove that the association between scores on Factor III and the temple-centered concentric zone system is quite weak.

Thus, again, one may conclude that we are not justified in rejecting a null research hypothesis. There is no meaningful relationship between score on Factor III and the temple-centered

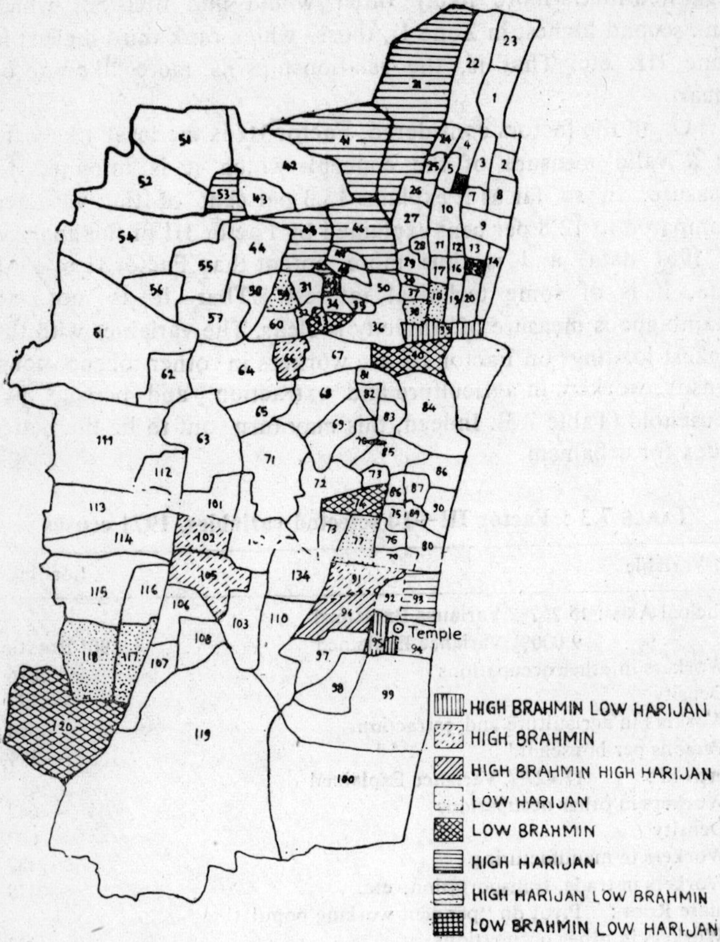

FIG. 7.3 : *Madras—Ethnicity*

concentric zone system.

Since it is difficult to make clear substantive inferences from this lack of relationship because of the questionable validity of Factor III, an analysis may be extended by examining Factor IV as a more likely candidate for an ethnicity factor. It is, among the other factors remaining in an exhaustive analysis, of particular interest because the variables with the highest and second highest loadings are per cent scheduled tribe and per cent scheduled caste, the only two ethnic group characteristics included in the Indian census. Table 7.4 reports the details of this factor and a partial test of the hypothesis that it is associated with the temple-centered concentric zone system. For this test, the divisions which were sampled for the household survey have been chosen and classified according to score on Factor IV. By the implied hypothesis, divisions in Social Area I, the area of lowest rank on the factor and thus the one containing the smallest percentage of scheduled caste and tribe members, will be found in the zone closest to the temple. If this hypothesis is interpreted in the strict sense of Hypothesis 3, then the divisions in Social Area II, the area of second rank on Factor IV, will lie in the zone most distant from the temple, etc. If the alternative (linear) interpretation is applied, then the order of the social areas will be identical to the order of the zones in which they are found. It is seen that there is a fair degree of association between expectations and results in the case of the latter interpretation, a degree great enough to lead one to suggest that Factor IV is the better index of ethnicity. This would, in turn, imply that a factor which explains a relatively large percentage of the variance does not necessarily relate to an important characteristic of the population (important, that is, in terms of ecological conventions).

TABLE 7.4 : **Factor IV—14 selected variables, 1971 census**

Principal Axis : 10.240 per cent variance explained
Centroid : 9.720 per cent variance explained

Variable	Loading		Square root
	Principal axis	Centroid	
Per cent scheduled tribe	.683	.539	.985
Per cent scheduled caste	.582	.492	.333
Workers in trade, transport, etc.	—.451	—.445	—.105
Workers in manufacturing	—.312	—.297	—.044
Workers in household industry	.256	.218	.368

General conclusions

The most obvious conclusions which can be drawn from the test of the three hypotheses is that there is no basis for suggesting a relationship between concentric zone or sectoral formation and social area based upon demographic characteristics. As has been mentioned in relation to specific instances, however, it is difficult to infer from this that Madras displays no regular ecological structure. It is suspected that this is the case but there is lack of firm evidence. In a less formal way, first-hand observations in Madras support this suspicion. Within a radius of perhaps one or two kilometres from nearly any point in the city one can find neighbourhoods encompassing the entire range of social characteristics, socio-economic, family structural, or ethnic-group related. Why this is the case for Madras but is apparently not so for other cities submitted to ecological analysis is a complex question and one for which one lacks formal evidence. But it is a question which is important enough to warrant a few words.

In the first place, as has been implied, there is reason to doubt the validity of certain conclusions drawn from the study of the ecological structure of other cities, especially those in third-world societies. As a result, it is difficult to differentiate between those things which are characteristic of Madras in particular and those which only seem to be so. Notwithstanding this, we would suggest that the ecological structuring of Madras is fundamentally that of a multiple nuclei form in a very special sense. It is likely that Madras (and perhaps any Indian city) represents an association of villages within which one might find the kinds of settlements and patterns closer to expectations but between which there is no clear pattern. This is not to suggest, then, that the three-setting principle which have been introduced is of no use. To the contrary, it might prove to be most useful. It is likely, however, that the unitary character of central tendency which these settings and the spirit of sociological studies constrain one to assume is inappropriate to the Indian situation. In this regard one may, in fact, no more than reiterate well established observations :

> It would appear impossible to recognize the "orthogenic" character of India's cities and villages and at the same time to think of comparing them separately with their "equivalents elsewhere".

THE ECOLOGICAL STRUCTURE OF MADRAS—SOCIAL STRUCTURAL

Supplementary information was gathered from a sample of 2,000 households within the municipal boundaries of Madras City, as has been noted, because of the difficulties encountered in attempting to relate findings based solely upon census materials to traditional ecological expectations. The purpose of this section is to report some of the results of this survey. The form of this presentation is that of a replication of some of the investigations discussed in the previous section.

There were several variables related to items on the interview schedule which are likely indicators of socio-economic rank. Eight of these were chosen for the purpose of constructing a social rank index : total household income, years of education (male head of household), and percentage of heads of household in each of six occupational categories : unskilled-labour, semi-skilled labour, skilled labour, semi-professional, professional and managerial, retired and unemployed.

The principal axis factor analysis of the correlation coefficients between these variables was performed for the first three factors and factor scores were based upon an orthogonal rotation. Table 7.5 details the composition of the three factors produced, although scores were based only upon the first of these, which

TABLE 7.5 : **Social rank index—household interview survey**

Factor I : 52.88 per cent variance explained
Factor II : 21.47 per cent variance explained
Factor III : 11.36 per cent variance explained

Variable	Loading		
	Factor I	Factor II	Factor III
Total household income	.709	—.357	.492
Years of education (male)	.853	.252	.239
Per cent in occupational category			
I	—.701	.570	.026
II	.286	.874	—.005
III	.730	—.480	—.184
IV	.907	.197	—.037
V	.875	.411	.060
R and U	.553	.001	—.755

explains more than 52 per cent of the variance. Generally, a division with a high score on this factor is likely to be characterized by a high mean income, a high number of years of education, a low percentage of the population classified as unskilled workers, an intermediate percentage of semi-skilled workers, and a high percentage of workers in the categories of higher rank.

A comparison between the actual bazaar-centered zone in which each division can be found and the zone which would be expected for it given its score on the social rank factor—and granting the truth of Research Hypothesis 1—shows that there is not a high degree of agreement. The actual form of the structuring is as difficult to interpret as was the case for social areas based upon strict demographic variables.

Family structure index

Of the three indices under consideration in this section, the family structure index is the most interesting in its own right. The information contained in Table 7.6 leads to some interesting questions with regard to the inner structure of these factors.

TABLE 7.6 : **Family structure index—household interview survey**

Factor I : 57.98 per cent variance explained
Factor II : 17.89 per cent variance explained
Factor III : 10.90 per cent variance explained

Variable	Loading Factor I	Factor II	Factor III
Persons per household	—.696	—.407	—.589
Years of education (female)	.488	—.826	.225
Age of male head of household	—.881	.064	.189
Years at present address	—.897	—.021	.126
Years in Madras	—.773	—.204	.311

The first factor produced by a principal axis analysis, one which explains nearly 58 per cent of the variance, is the strongest factor produced in the entire study. In choosing the variables for this index, the suggestion of Shevky and Williams (1949) was followed. The expectation was that an area at one extreme of this index would be characterized by a low number of persons per household, a high mean number of years of education for the female head of the household, a low mean age of male head of household, a low number of years at the present address, and (and here is where an empirical discrepancy arises) a high number of years in Madras. In this way, there would be a reasonable mathematical model of a neolocal, upwardly mobile, nuclear, urban family ; that is, there would be an index for the ideal North American family structure.

In fact, as can be judged from Table 7.6, the expectations were satisfactorily fulfilled, with one exception. For, according to the loadings on Factor I, such an area would be characterized, instead, by a low mean number of years in Madras—an apparently anti-urban characteristic. Since this variable is highly correlated with both mean number of years in Madras and mean age of male head of household, the negative sign of its loading is understandable. Its absolute value (.773) is not understandable by these correlations alone. In addition, there is external evidence that the nativity index correlates uncertainly with number of persons per household and, if anything, negatively with female education. This provides further warrant to question the expectation that prolonged urban residence is clearly associated with the other characteristics of a high-ranking division. If these considerations are sound, then another example of the ambiguity of the word "urban" when applied to a third-world city has been provided.

By classifying the divisions sampled according to score on the first factor of the family structure index it was possible to judge the extent to which the zone in which each is found compares with the zone relating to the fortress-centered concentric zone system. The comparison gives clear indication of no relationship. Thus, the decision made on the basis of census data may be reaffirmed.

Ethnicity index

Whereas the "discovered" ethnicity index based upon score on Factor III of an analysis of census data proved to be the most troublesome with regard to validity, the constructed ethnicity index based upon interview survey data which will be now discussed, to the contrary, likely to be among the most interesting products of this study. This index includes information on 25 variables, 20 of which relate to caste composition of the 40 divisions from which the sample was drawn. Table 7.7 reports the structure of the first three factors produced in a principal components analysis of this information.

The first factor, from which factor scores were calculated explains slightly more than 17 per cent of the variance. Divisions

TABLE 7.7 : **Ethnicity index—household interview survey**

Factor I: 17.082 per cent variance explained
Factor II: 10.307 per cent variance explained
Factor III: 9.820 per cent variance explained

Variable		Loading Factor I	Factor II	Factor III
Per cent born outside Tamil Nadu :				
	M	—.664	.214	.438
	F	.726	*	—.268
Per cent born outside Madras :				
	M	.766	—.184	—.220
	F	.560	.508	.279
Mean length of residence (Madras) :				
	M	.658	.429	.407
Per cent in caste :	Brahmin	.572	—.237	—.142
	Harijan	—.509	.414	—.273
	Christian	—.527	.457	*
	Chettiar	.173	—.348	.466
	Muslim	—.429	.390	—.133
	Naicker	.126	—.360	—.537
	Mudalier	—.164	—.241	.242
	Naidu	—.476	—.248	.311
	Vaisya	.523	.482	.293
	Pillai	*	—.235	—.210
	Jain	.543	.407	.365
	Rao	*	*	.265
	Gramani	*	*	—.493
	Sowarashtra	—.174	—.312	.555
	Achari	*	—.353	.347
	Nair	.164	—.268	.262
	Yadavha	.240	.260	—.123
	Vanier	—.246	.165	—.205
	Valluran	.142	—.476	.403
	Chetty	*	—.353	—.117

* Lies between —.09 and + .09.

which score high on this factor are likely to be characterized by a high percentage of heads of household, both male and famale born outside Madras and a high percentage of heads of household born outside Tamil Nadu. The implications of this, especially when considered in combination with the caste loadings, are especially interesting, for a division scoring high on this factor can also be characterized by a high percentage of the population belonging to what are generally considered to be high caste groups and a low concentration of low caste (e.g., Harijan) groups. By virtue of this, one more index in which high social rank is associated with non-urban (as defined by place of birth) characteristics has been found.

Another interesting feature of this index has to do with the relative values of the loadings for each caste group considered. In this instance, it may be the skeleton of an interval-level measure of caste ! As an operational fact of daily life in India there exists a widespread and precise understanding of the relative rank of the prestige of castes which contribute heavily to the life-chances of individuals and groups. However, when the sociologist deals with caste as a variable, he is generally constrained to deal with it as a nominal variable and to treat only percentages at higher levels of measurement ; that is, with the exception of the four traditional Vedic Varna which can be treated as ordinal. This necessity often produces results which are significantly uninteresting compared to what "everybody knows" about the structure of the caste system. But, treating loadings on this factor as a measure of how strongly and in what direction the caste is associated with overall ethnic characteristics face validity at the interval level holds. With further refinement, an index such as this can be employed in cross-validity checks. The reader may judge the extent to which relative values of these loadings agree with his understanding of the caste system, for, it may be contended, if this index points to a latent phenomenon, that phenomenon is caste itself.

A comparison between score on the ethnicity factor and the temple-centered concentric zone system shows that the relationship is weak at best. However, if the relationship between hypothetical temple-centered zone and score on the ethnicity factor were to be treated as linear, the situation is changed and relationship appears significantly stronger. This change is productive of the most promising result of this study. For, clearly, there is a tendency for groups scoring high on the ethnicity index to live close to the temple and for those scoring low to live a considerable distance away. This tendency has two germane implications. First, a more stable basis for the inclusion of the temple as a major institution-related setting has been provided. Of the three settings, it seems to play the strongest role in the ecological structure. The second implication involves the first bit of empirical evidence with which one may question the application of the Park and Burgess model to ethnic characteristics. That is, if the formula (1 5 4 3 2) applies to the relationship between rank and location in any institutional dimension (for Madras at least) then it is equally likely that the formula (1 2 3 4 5) applies to ethnic characteristics.

In general, this supporting material from the household survey has been just that supporting. None of the conclusions drawn in preceding discussion have, by the use of this information, been radically altered. It is fairly clear that neither socio-economic rank nor family structure characteristics display a concentric zone system of any form and that ethnic groups characteristics might, provisionally, be considered to be so configured in the ecological structure only in a way different from hypothetical (although not, perhaps, common sense) expectations.

IMPLICATIONS FOR COMPARATIVE URBAN ECOLOGY

Throughout this chapter an argument concerning an alternative to the now weakened hypothetical ecological structure of Madras has been adumbrated. This discussion is concluded by cataloging a few of the insights coming from this study which may be helpful in framing a more exact alternative.

Of the three (or four) census-based factors which have been produced in ecological studies of cities in western countries, only social rank is found in Madras. Further, regardless of the validity of factors, the model industrial metropolis provides no better explanation of geographic distribution of social characteristics than does chance alone. While it is clear from a demographic and historical standpoint that India and Indian cities have changed in the past several decades, it would be impossible to infer from the study that the ecological structure of Madras is modernizing. The more justified inferences which can be drawn are that both modern and traditional elements are present in Madras' residence patterns and that evidence has yet to support either clear divergence or convergence with respect to the model industrial structure.

Although the institution-related settings do not appear to be unambiguous centres of concentric zone systems, their importance in the history and daily life of the city is undeniable. In general, Madras lacks the large-scale natural area formation associated with the concentric zone or sectoral models. Between (and within) tract variance is high. Regularities exist, of course, but they tend to occur on a relatively small scale. To the extent that the ecological structure of Madras does share features with the models, this structure appears to be organized in a microcosm.

Natural area formation depends crucially on residential segregation of social strata, and residential segregation of this type is not characteristic of Madras. Certainly this is the case with the more prominent and dramatic aspects of residence (houseless persons, squatter settlements, and slums) which produce apparently strange spatial juxtaposition of social groups. This also appears to be a feature of the less visible but more deeply entrenched aspects of neighbourhood formation.

These points indicate that a principle which is different from that assumed by the models is present in Madras. The fact that the temple provides the most meaningful focus for the geographic distribution of social characteristics, and that these characteristics are caste associated, suggests the nature of this principle.

Ecological structure articulates characteristics of social structure in a space-bound context. The residential segregation assumed by the models reflects a general democracy of observable behaviour, a relative lack of overt signs of rank. In the industrialized cities upon which these models are based, Chicago, Washington D.C., Boston, and so on, other than race and ethnicity (characteristics whose distribution in Madras most closely agrees with expectations), location of residence is the prime symbol of the rank of families. Beshers (1962) has discussed this fact in pointing out that few other acquired aspects of life in the United States are as fixed and as highly visible as one's address.

The relatively traditional forms of social ranking in India do provide for overt rank identification : dress, dialect, manners, name, and, after all, skin colour. Residence has not yet come to serve the symbolic function which is still performed by these social features. The lack of clear residence patterning in Madras, then, reflects the overwhelmingly rural and traditional character of the region and the nation.

For the next several years Indian urban sociology will have before it the task of integrating the recent census with urban ethnography and survey research findings and, in turn, relating these data to social theory. The sociologist whose philosophy of science cannot tolerate empirical contradiction may find this challenging task impossible, for, however one chooses to approach the social structure of contemporary India, he finds ambiguity and paradox. Madras is no exception.

In Madras, a city of $2\frac{1}{2}$ million people, the adaptation of the ecological structure to the forces of tradition and modernity seems to have produced an organization of urban villages in a most literal sense. N.K. Bose has suggested that this kind of paradox follows from the premature nature of cities in India :

> These cities have arisen out of phase with history ; they have appeared in the setting of the traditional agricultural economy in advance of the industrial revolution that is supposed to beget the metropolis (Bose 1965).

Thus residence patterns in Madras display a high degree of heterogeneity within relatively small areas because, as with other aspects of urban India, the social structure is only nominally urban in the Western sense. If an open class system is universally produced in the process of industrial revolution, so, too, is the regular ecological structure which reflects such a stratification system. India's traditional system endures and has endured with essential integrity in the face of many "invasions". As recently as 1971 this system persists in the geographic distribution of households in her fourth largest city.

Future research may indicate that the ecological structure of Madras will become comprehensible in the light of the models employed in this study. Our results indicate, however, that until such time, analysis of block rather than tract level data is more than a methodological refinement, for until natural area formation begins to occur on a larger scale, the distinctive features of the ecology of this and other Indian cities will be overlooked.

The study of urban ecology in India has barely begun. Much research is needed, especially historical and time series types of research. As Western models prove deficient, there is an opportunity for a unique adventure of discovery into the ways in which India's complex and fascinating social system solves the problem of distributing space among hundreds of millions of people while simultaneously making its adjustment to the conflicts of tradition and modernity.

EIGHT

JAY A. WEINSTEIN AND V.K. PILLAI

AHMEDABAD
An Ecological Perspective*

INTRODUCTION

RECENT research on Indian cities employing the ecological perspective has focused on the social composition of urban neighbourhoods as an indicator of modernity. Some of this work suggests that the modern configuration of family structure, ethnicity and social rank (and, later, urbanism), first observed in the United States by Shevky and his associates (Shevky and Williams 1949, Shevky and Bell 1955, Bell 1955), is coming to dominate the structure and spatial distribution (as first elaborated by Anderson and Egeland 1961) of the populations of Calcutta, Poona, Madras and Ahmedabad (Berry and Rees 1969, Berry and Spodek 1971). While various revisions of the Shevky model have been suggested (Anderson and Bean 1961, Van Arsdol, Cammilleri and Schmid 1958, Schwirian and Matre 1969), its basic elements appear from this research to be discernible in the social geography of large Indian cities as well as elsewhere in the third world (Latif 1974).

Other studies of Indian cities, however, have indicated that the influence of traditional/local institutions remains strong (Weinstein 1974, 1976b ; Mehta 1968, 1969). While this research

* Reprinted from *Third World Planning Review*, Vol. 1, No. 2, Winter 1978, by permission of the author and editor.

recognizes both the presence of modern elements and changes in the relationship between modern and traditional forms, it also suggests that modernization, in a cumulative sense, has not yet been demònstrated in ecological studies. Instead of a clear tendency toward domination by modern systems of social ranking, etc., and the associated modern spatial configurations, these cities appear to be experiencing an increasingly complex interpenetration of modern and non-modern sociographic systems (Weinstein and McNulty 1979).

In this paper we examine the case of Ahmedabad principally as a descriptive exercise ; yet our findings make possible some limited inferences about the presence or absence of a modernization trend.* The physical features of Ahmedabad clearly reflect a dramatic interpenetration of old and new, native and foreign. Particularly striking is the contrast between the old walled city** and the rest of the metropolitan area which grew up around the walls. The old city has an area of about 2.1 square miles and is bordered on the west by the Sabarmati river. The wall originally had seven gates, or *dharwajas*, where settlements and communal activity centered. Today the wall is in ruins and most of the *dharwajas* are only traffic intersections. The metropolitan area now encompasses 35.9 square miles. Its social geography is dominated by textile mills, shops, and offices rather than by the fortress or the bazaars, temples and mosques within its vestigal walls.

Yet, as the city has grown, its physical features have been transformed not so much by replacement as by accretion. Contemporary Ahmedabad consists of distinct natural areas which vary in age, land use, and location in relation to the wall. This

* The data reported in this paper are taken from the 1961 and 1971 Government of India Census Reports on Ahmedabad Municipal Corporation and from a 1975-76 social survey of over two-thousand households in the city. These data are employed in an examination of the current social structure and geographic distribution of Ahmedabad's population.

** Walled cities, of both Muslim and Hindu rulers, were common in the region. As Spodek observes ; "A combination of high rural taxation which antagonized farmers and a policy of predatory militarism toward neighbours led rulers in western India, for example, to build their cities as remote fortresses for protection rather than as open market places for trade" (1975).

natural area formation reflects the broad sequence of eras in Ahmedabad's nearly six-hundred years of history : Mughal rule, Hindu conquest, British colonization and independence. By focusing on these natural areas in an examination of Ahmedabad's contemporary ecological structure, we may be better able to judge the extent to which the historical and physical characteristics of the city have a counterpart in its social system. Are the walled city and the other natural areas distinct *social* areas ? Is the traditional character of much of the architecture on the east bank of the river indicative of the social rank, family structure and ethnic character of the population ? Or, does the British Colonial style somehow remain dominant even in social relations ? Or, finally, is the le Corbusier-inspired style, which is apparent on the west bank and even at points in the old city, the true indicator of Ahmedabad's present social structure ?

While the picture that emerges from this examination of Ahmedabad is a complex one, it tells of the persistence of traditional forms in a modern third world city.* While the physical wall exists only in vestiges, in some respects its social functions remain relatively intact. While the mills dominate the skyline, and suburban developments grow in size and number each year, their social effects have, in some ways, not yet come to dominate the traditional influence of caste, language, region and family.

The metaphor of the wall and the mills may serve to remind students and planners of social change and urbanization in India (and elsewhere) that traditional society is not merely passive toward modernization. Though it appears that modern norms and technologies can easily come to replace traditional-local ways, such appearance often hides a deeper structure. This structure is partly revealed when, in cases like Ahmedabad, the effects of physical landmarks and historical processes are observed in present neighbourhood settlement and land use patterns.

* As Kenneth Gillian has noted :

"Unlike Bombay, Calcutta, Madras and Kanpur, Ahmedabad was not a creation of the British but a city, which while remaining true to itself, successfully adopted (the) industrial age carrying over commercial and industrial skills and patterns of traditional social organization. In no great city of India can the continuity of past and present be seen as clearly as in Ahmedabad" (Shah 1976).

AHMEDABAD'S NATURAL AREAS

Ahmedabad is the seventh largest city in India (after Calcutta, Bombay, Delhi, Madras, Bangalore and Hyderabad). It is one of the most densely populated cities in the world, with a 1971 popu-tion of over 1.5 million, or 41,907 persons per square mile (compared with Macao which has about 44,000 persons per square mile). The city was established in 1411, on the site of an ancient Hindu settlement, by the Mughal king Ahmed Shah. Among the first structures built were a royal palace, a fort wall along the city boundaries, a main street connecting the palace to the main gates of the fort wall, and a mosque located between the main gate and the royal palace. The fortress was completed in 1448, on a spot with a commanding view of the Sabarmati river, which flows north-south through the city.

During the first two centuries of its history, Ahmedabad flourished as a manufacturing and marketing centre. Toward the middle of the eighteenth century, trade began to decline (Choskey 1968). During this period, the power of the Mughals diminished and the Marattas—conquerors of the region—took over administration of the city in 1738. Looting and terrorism broke out in the wake of tension between the new Hindu rulers and the now-subjugated Muslims. The remaining commercial families fled ; and solitude, poverty and desolation struck the city. Ironically, this period of civil disorder (lasting until 1817) also gave rise to the unique and exquisite dwelling patterns which still characterize the old city today. Fear of arson and looting caused the residents to construct closed apartment blocks, three to four stories tall, which were designed as minifortresses within the main fortress walls. The principal type of these structures is the *pol*.

Each *pol* is a closed loop with an open community space, called a *chowk*, at either end or in the middle of the loop. Houses are built with their side walls flush with those of adjacent houses, in part for security and in part as insulation against the intense heat of the summer sun.* The looped streets were formed to

* One can still travel nearly from one end of the walled city to the other by stepping from rooftop to rooftop. At one time an extensive system of underground tunnels connected the *pols*.

protect privacy, but they have often also served as ghetto boundaries segregating ethnic groups from one another. Originally, toward the end of the Mughal period, the *pols* were used as quarters for the king's courtiers and army personnel. With the decline of Mughal power, the merchant and money-lending communities (e.g., Hindu and Jain Baniyas) moved into the *pols*. In the post-independence period, the economically mobile and upper-income groups have been moving out of the *pols* to the far eastern suburbs and the west bank, although many of these wealthy families retain title. The *pols* are now occupied partly by lower income and/or lower caste groups. In-dwelling densities are extremely high because of continuous subdividing and subletting of residential space.

Ahmedabad came under British rule in 1817 (compared with the coastal cities of Calcutta, Madras and Bombay which were established in the mid-seventeenth century). Commercial activity was revitalized under the British ; however, as Gillian (1968) observes, "the British presence was never obtrusive in Ahmedabad as it was in Calcutta or the other cities they had built as their own, or had made into military stations like Poona." Civil peace and the advent of industrialization in the ancient textile sector led to a dramatic period of economic reconstruction. The Gujarati Baniyas, who were traditionally money-lenders, assumed a position of economic and political dominance in the local community. Under the proto-capitalist entrepreneurship of these families, Ahmedabad became a centre of England's modern textile industry.

The first full-scale textile mill was established in the eastern section of the city, just outside of the wall.

> By 1891, thirty years after the opening of the first mill, Ahmedabad had a total of nine mills. The Baniya community had entered the industrial field. Modern transportation and communication facilities were introduced and education spread. Institutions for financing and managing the mills had been developed. Channels for obtaining machinery and for recruiting personnel from England had been opened (Spodek 1965).

By 1900, Ahmedabad had been thoroughly "Manchesterized".*

The expansion of the city's industrial base resulted in large-scale in-migration of the rural population, especially from the west and north. As of 1971, approximately half of Ahmedabad's population was born outside the city. Today, about twenty per cent of the population lives in workers barracks and tenements in the mill district beyond the walls. Poor sanitary conditions, constant smoke and cotton dust pouring from the brick chimneys, and astoundingly high rate of tuberculosis,** continue to reflect the darker side of Ahmedabad's industrial prosperity.

The west bank of the river was settled by wealthy money lenders and businessmen during the Maratta period. They developed large unoccupied tracts of land into several small market centres, or *puras*. After industrialization, however, the revitalization of the economy on the east bank caused the *puras* to diminish in importance. Eventually they became residential areas for the middle and upper classes. There are a few squatter settlements in this part of the city today, including the extensive Liberty Hutment along University Road in Navrangpura (an old *pura*, now part of Ahmedabad City) ; however the squatter population on the west bank is small compared to that found in the industrial areas in the eastern part of the city.

Industrial activity has, in recent years, been transformed because of Government regulations. In 1947, the Government of India introduced new regulations to control the expansion of the textile industry and to give impetus to the handloom sector.*** These restrictions did, perhaps, slow the half-century of expansion

* At the same time, as Spodek and other historians of the British Raj in Gujarat have pointed out, Manchester was being "Ahmedabadized" through the import of technology and design principles from India. Thus pairs of cities such as Ahmedabad and Manchester have played a very special role in the growth of the modern world-system (see, e.g., Wallerstein 1974).

** One local physician, a lung specialist trained in New York, estimates tuberculosis rates to be as high as 500 in 1000 children and older adults.

*** Gujarat is Mahatma Gandhi's home state, and his Ashram is in the northwest section of Ahmedabad. The movement for handloom production, and other aspects of his political programme, were deeply rooted in Gujarati traditions and, in principle at least, were widely accepted there (Spodek 1971).

in textile production, but they also lead to the diversification of the industrial base, principally in chemicals and related products. While textiles still dominate Ahmedabad's manufacturing and commercial sectors, the economy has become increasingly dependent on value added from non-textile activity.

The present city of Ahmedabad developed through the addition of land and through population growth in relatively distinct historical stages (Shah 1976). The old city, within the fortress walls, dates from the fifteenth century. The *pols* and the old *pura* organization on the west bank are associated with the mid eighteenth century. The industrial section just outside the walls, with its workers quarters and squatter settlements, was established early in the nineteenth century. The development of residential areas in the far-eastern and western sections followed the economic restoration of the city. The establishment of administrative and commercial centres on the west bank is even more recent. Today, the west bank continues to experience the most rapid growth in population (a 94.6% increase between 1961 and 1971) and in commercial development.

Present-day Ahmedabad consists of four large natural areas which vary in age and in distance from the original site of Ahmed Shah's palace in a characteristic concentric zone manner, as shown in Fig. 8.1. Area I, the walled city has a total population (1971) of 480,735 in 2.12 square miles. The mean ward density in the walled city is more than 225,000 persons per square mile (about the same as on Hong Kong Island).

Natural Area II, the near-peripheral area, is the mill section of the city. This area is bounded on the west, southwest and northwest by the wall and on the east by natural Area III. It has a total population of 289,439 in 7.51 square miles; an areal density of 38,540 persons per square mile.

Area III is the far-peripheral zone, bounded by the mill section and by the municipal limits of Ahmedabad. This area contains a scattering of mills, Ahmedabad General Hospital, The Ahmedabad Zoo and Kankaria Tank (an artificial lake) to the south, Malek Saban Stadium to the west, and large residential sections which now extend beyond the city limits. Its total population is 456,284 in 10.64 square miles, an areal density of 42,884 persons per square mile. The population of Area III increased by 31.8% between 1961 and 1971.

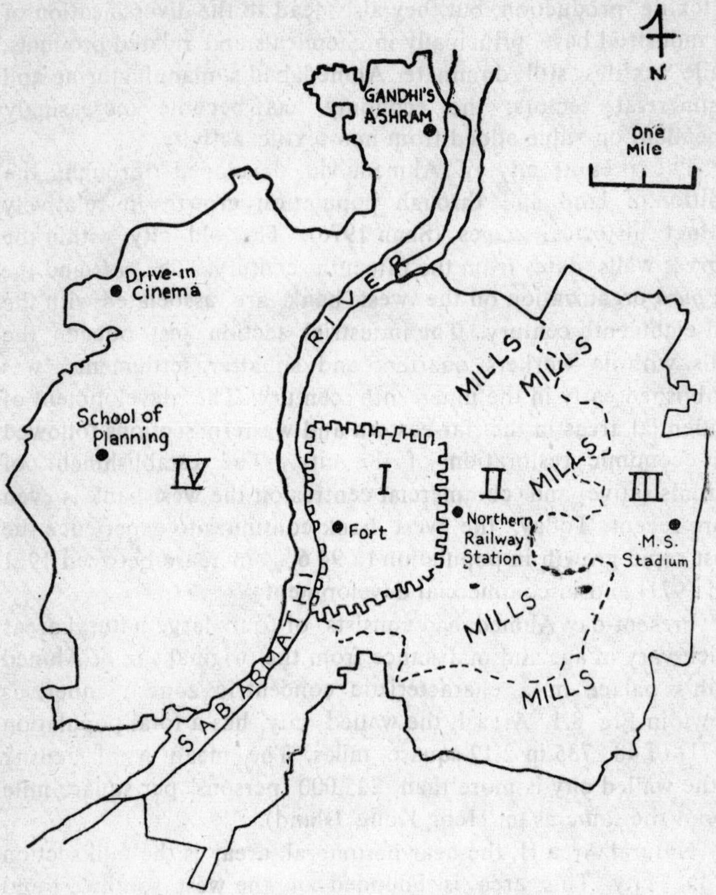

FIG. 8.1 : *The natural areas of Ahmedabad*

Area IV consists of Ellis Bridge/Navrangpura, Ahmedabad's largest ward in both area and population size with a total population of 222,954 in 13.7 square miles and Sabarmati ward. The total population of area IV is 268,905 in 15.63 square miles : a density of 17,204 persons per square mile, or about 1/13 that of the walled city. This area was the site of several small villages scattered along the river bank and inland from north to south. Today, juxtaposed on these *puras*, is Ahmedabad's principal suburban development : apartment blocks, bungalows, small commercial districts, the State University, several other leading colleges and research institutes (including the Indian Institute of Management, Sardar Patel Institute, and the School of Planning) a drive-in cinema and, to the north, Gandhi's Ashram, now a national monument.

AHMEDABAD'S PRESENT SOCIAL STRUCTURE

Census data

In this and the following section the social characteristics of contemporary Ahmedabad and its natural areas* are examined. In this section the attention is on the social structure of the general population, independent of differences between natural areas or wards. In the following section the focus is on the extent to which Ahmedabad's natural areas correspond to social areas.

The history and physical characteristics of the city point to a complex process of accommodation and accretion. With the use of census and survey data it is possible to investigate whether and how this process is reflected in Ahmedabad's present social structure. Table 8.1 lists twelve selected census variables and the value on these variables for the city and the four natural areas. As of 1971, 47 per cent of the work force was employed in manufacturing**—a reduction of about 7.0 per cent since 1961. The second most frequently reported occupations are trade, commerce, and other (generally clerical) services (43.7% in 1971).

* Census and survey data have also been collected and analyzed at block and "division" level. While we report only ward—and natural area-level findings in this paper, results from the finer levels of aggregation are likely to be more revealing. It should be stressed that intra-ward variances on most of the variables used in this study are high. For further comments on the advisability of using small areal units see Schwirian, 1971; Weinstein, 1974 and Pampel and Choldin, 1978.

** In the census of Ahmedabad, household manufacturing is classed with manufacturing in general. This is unfortunate since these are essentially two different types of manufacturing activity, one associated with traditional and rural social structure. We have followed the option of treating this variable as a unidimensional indicator of modernity. Here, however, is one more reason for being sceptical about conclusions drawn from the analysis of census data alone; for other reasons see Weinstein, 1976b.

TABLE 8.1 : **Twelve selected census variables—city and natural areas (wardwise) means (1961 and 1971)**

	Persons per house-hold	Per cent sche-duled caste	Per cent sche-duled tribe	Per cent literate females	Per cent working females	Per cent working popula-tion	Percentages of workers employed in					Per cent literate popula-tion
							Agricul-ture and extrac-tion	Manufac-turing	Trade and com-merce	Trans-port	Other ser-vices	
City												
1961	5.45	9.3	0.4	34.7	7.5	30.1	0.7	54.3	18.8	6.8	23.0	52.7
1971	5.49	10.4	0.5	38.8	7.2	28.2	1.1	47.0	23.2	8.2	20.5	61.9
Area I												
1961	5.59	3.7	0.3	40.0	6.2	27.1	0.2	45.7	26.3	4.5	23.8	64.5
1971	5.88	3.1	0.3	42.5	6.3	26.4	0.5	42.4	30.4	5.6	21.1	70.2
Area II												
1961	4.80	23.9	0.7	27.2	10.4	37.0	1.4	53.7	12.2	6.3	26.4	42.1
1971	5.18	28.4	0.9	32.9	9.9	31.7	1.3	48.3	17.9	11.6	20.8	48.8
Area III												
1961	5.83	7.4	0.3	31.5	5.5	32.7	0.4	66.5	11.3	6.4	16.3	51.0
1971	5.15	9.3	0.4	36.6	6.3	28.9	1.0	59.1	15.7	7.6	16.6	58.3
Area IV												
1961	4.91	9.0	0.7	35.8	15.5	18.5	3.1	63.1	19.4	*	*	*
1971	5.22	8.7	1.2	40.2	8.3	27.2	4.9	24.2	21.9	17.3	31.7	61.6

*Missing data

Just over 28 per cent of Ahmedabad's total population is in the work force, with just over 7.0 per cent of the city's women reported as gainfully employed. General and female literacy rates both increased by more than 1.0 per cent per year during the period (with male literacy increasing considerably more rapidly than female literacy). The percentages of the population in scheduled castes and tribes both increased slightly between 1961 and 1971.

A test of the degree of modernization in the structural relationships among census of India variables has been developed by Berry and Spodek (1971) and revised by Weinstein (1976b). According to this approach, a principal axis, orthogonally rotated, factor analysis of these variables will indicate modernity to the extent that it approximates the model shown in Table 8.2. This

TABLE 8.2 : **Hypothetical factor structure for Ahmedabad**

Variable type	Variable (sign of loading)	Expected loadings on factors			
		I	II	III	Others
Social rank	Per cent literate population (+)	H	L	L	L
	Per cent literate females (+)	H	L	L	L
	Per cent working population (+)	H	L	L	L
Family structure (Urbanization)	Per cent working females (+)	L	H	L	L
	Persons per household (—)	L	H	L	L
	Per cent workers in agriculture (—)	L	H	L	L
	Per cent workers in manufacturing (+)	L	H	L	L
	Per cent workers in trade and commerce (+)	L	H	L	L
Ethnicity	Per cent in scheduled castes (—)	L	L	H	L
	Per cent in scheduled tribes (—)	L	L	H	L
Others	Per cent working in transport	L	L	L	H
	Per cent working in other services	L	L	L	H

Based on Bell 1955, Berry and Spodek 1971, Weinstein 1976c.
"H"=High expected loading.
"L"=Low expected loading.

may be compared to the actual factor structures as observed in 1961 and 1971, as shown in Tables 8.3 and 8.4.

As in the case of Madras (Weinstein 1976c), the comparison between model and actual factor structures reveals a complex mixture of expected and discrepant findings. The variables with the highest loadings on factor I of the 1961 analysis (Table 8.2), a hypothesized social rank factor, are : per cent working females (+.896), workers in agriculture and extraction (+.741) and per cent in scheduled tribes (+.583) and scheduled castes (+.536). The first two of these variables are hypothesized to occur on the second, family structure/urbanization, factor and the last two on the third, ethnicity factor. The variables with expected high loadings, general literacy, female literacy and working population

TABLE 8.3 : **Factor analysis of twelve selected census variables (1961)**

Variables	Factor I 18.88% Var. Exp. loading	Factor II 18.37% Var. Exp. loading	Factor III 17.36% Var. Exp. loading	Factor IV 13.28% Var. Exp. loading
Persons per household	—.206	—.064	.883	—.132
Per cent scheduled caste	.536	—.420	—.376	—.158
Per cent scheduled tribe	.583	—.008	—.220	.074
Per cent literate	—.469	.574	.640	—.047
Per cent literate females	—.249	.358	.875	—.146
Per cent working females	.896	—.126	.047	.232
Per cent working population	.245	.123	—.936	—.280
Per cent working in agriculture and extraction	.742	.015	—.011	.238
Per cent working in manufacturing	.167	—.935	.024	.253
Per cent working in trade and commerce	—.430	.631	.420	.163
Per cent working in transport	.339	—.989	.083	.879
Per cent working in other services	.377	.891	—.094	.066

did exhibit moderately high values, but all three are more highly correlated with other factors in 1961. Thus, while the low literacy rates observed on this factor do correspond to expectations, it would be difficult to characterize it as an unambiguous social rank dimension. Instead, factor I—1961 essentially displays a combination of traditional (even rural) occupational and ethnic group characteristics.

Factor II of the 1961 analysis shows high (absolute) loadings for percentage working in manufacturing (—.935), percentage working in other services (+.891), percentage in trade and commerce (+.631), and general literacy (+.574). These and the other loadings point clearly to a commercial/upper social rank configuration in which the absence of workers in manufacturing is strongly associated with literacy, the absence of scheduled castes and high percentages in non-manual trades. Variables expected to have high loadings on this factor, female employment, persons per household and per cent workers in agriculture, are

TABLE 8.4 : **Factor analysis of twelve selected census variables (1971)**

Variables	Factor I 37.9% Var. Exp. loading	Factor II 21.2% Var. Exp. loading	Factor III 18.1% Var. Exp. loading	Factor IV 7.5% Var. Exp. loading
Persons per household	—.593	.976	—.251	—.023
Per cent scheduled caste	.344	—.450	—.025	.664
Per cent scheduled tribe	.413	.085	.460	.170
Per cent literate	—.616	.604	—.073	—.334
Per cent literate females	—.854	.441	.101	—.238
Per cent working females	.036	—.113	.342	.917
Per cent working population	.923	.094	—.143	.164
Per cent working in agriculture and extraction	.132	—.060	.885	.072
Per cent working in manufacturing	—.258	—.905	—.336	.070
Per cent working in trade and commerce	—.112	.907	—.249	—.249
Per cent working in transport	.782	.191	.242	—.066
Per cent working in other services	—.169	.482	.728	.336

only weakly (and negatively) associated. This pattern indicates not a family structure/urban-rural second factor, but rather something which looks very much like an urban, white-collar/urban, non-white-collar dimension.

Factor III—1961, a hypothesized ethnicity factor, displays a combination of high loadings for percentage working population

(—.936), female literacy (—.875), persons per household (—.883), and general literacy (+.640). In some respects these findings and the moderate loadings for percentages in scheduled castes and tribes indicate social rank. In other ways, however, this factor points to a dimension of Ahmedabad's social structure dominated at one extreme by an association between households which are large (reflecting high fertility and/or extended family structure), highly literate, in which there is typically one wage earner, and in which trade and commerce tend to be the most common occupations (loading +.420). These features tend to make factor III a combined white collar (like factor II) family structure/urbanization index. Factor IV, the residual factor, points to the structural segregation of transport workers and a, weak-to-moderate, presence of lower-middle social rank households.

The overall factor structure for 1961 displays the presistent presence of traditional and rural characteristics crosscutting those associated with more modern, urban and even class-like elements. Despite the serious problems of validity associated with the use of Indian Census data in this type of confirmatory factor analysis, it does seem clear enough that the expected three-dimensional configuration of social rank, family structure and ethnicity does *not* describe Ahmedabad's social structure, at least not according to 1961 census data.

In 1971 (Table 8.4), the first factor has high loadings for : per cent working population (+.923), per cent literate females (—.854), per cent working in transport (+.782) and per cent literate (—.616). Variables which had high loadings on this factor in 1961 have only moderate to low loadings in 1971. These changes between 1961 and 1971, combined with the moderate loadings on percentages in scheduled tribes (+.413) and castes (+.344), indicate a trend away from the domination of traditional elements and towards a less-ambiguous social rank dimension. It is a dimension which, at one extreme, is characterized by an association of households in which non-industrial, non-commercial employment is common, literacy rates are low and which contain a fair number of scheduled caste and tribe members.

The patterns observed in factor II for 1961 are present, and in some ways more pronounced, in this factor for 1971. A very high, negative loading on per cent working in manufacturing (—.904) combined with high (positive) loadings for per cent working in trade and commerce (+.907) and in other services (+.482) and high literacy (general, +.604 ; female, +.441) all strongly indicate an urban white collar/non-white collar dimension. Thus, in 1971, both factors I and II indicate social rank ; factor I in a general sense which combines traditional/rural and modern features, factor II in a sense which distinguishes on the basis of specifically modern, manual-nonmanual, traits.

Factor III—1971 seems, better than any of the other factors in either year, to indicate a pure urban-rural dimension. High positive loadings on per cent in agriculture and extraction (+.885), per cent in other services (—.728), percentage in scheduled tribes (+.459), and high percentages of working females are complemented by moderate negative loadings on percentages in urban occupations (manufacturing, —.336 ; trade and commerce, —.248). The aspects of this factor in the 1961 analysis which cause it to resemble factor II—1961 do not appear in 1971. Instead, factors II and III—1971 display a marked separation between urban social rankings and urban-rural differentiation. Factor IV—1971 tends to indicate a separate traditional social rank dimension ; with high positive loadings on per cent working females and per cent scheduled caste, and moderate negative loadings on literacy and per cent working in trade and commerce.

The overall factor structure for 1971 shows a clear convergence toward an ordered three-dimensional model. This is strongly indicated in the differences in the percentages of variance explained between 1961 and 1971. The 1961 percentages are very evenly distributed among the first four factors, whereas in 1971 an approximate 2 : 1 relationship exists between factors I and II and II and III. The eigenvalue for factor IV is below 1.0. We also see an increasing tendency for social rank to dominate yet, we still fail to observe a clear trend toward the hypothesized modern tripartite structure. Instead, we see a separation between general social rank (factor I), white collar-non-white collar differences (factor II), rural-urban differences (factor III), and traditional social rank (factor IV). Thus, while changes have occurred between 1961 and 1971, some point to increasing modernization (e.g., the changes in factor III and IV), some to increasing traditionality (e.g., the changes in factor II), and most to no clear trends.

Modernization of Ahmedabad's social structure cannot be discerned in neither census year for which data were analyzed. Like its physical structure, traditional, modern, urban, and rural elements all play a part in shaping the whole. In this city a distinct but familiar variant of what N.K. Bose described in his study of Calcutta can be observed :

> . . . the scene of a major confrontation between the *enduring* institutions of old India . . . caste, heritage and diversity of ethnic communities *and* presence of values arising from the process of urbanization (Bose 1965, p. 265 ; the author's emphases).

Survey data

As mentioned earlier, such conclusions must be considered to be tentative suggestions when they are based only on an analysis of census data. To help in achieving a better-grounded understanding of Ahmedabad's social structure, data were collected in a 1976 survey of households.* The variables were selected specifically to correspond to the U.S. Census variables upon which the model modern structure is based but which are not included in the Census of India study design (for a comparison between Indian and U.S. Census variables see Weinstein 1976c, pp. 37-38). Included in these survey variables are income (total household and male head's), education, rental, occupation in the sense of class rank (rather than sector-wise), age, birth place and caste.

Ahmedabad's mean household income, according to this

* Sampling was done on the basis of a three-stage, stratified-cluster (p.p.s.) —systematic procedure described in Weinstein, 1974. The variable "persons per-household" appears in both the census and survey data and can be used as a reliability check. The 1971 census reports the following persons per household for the city and the natural areas, respectively, for 1971 : City, 5.492 ; Area I, 5.884±.432 ; Area II, 5.179±.533 ; Area III, 5.145±.256 ; Area IV, 5.217±.316. Survey data show these values : City, 5.633 ; Area I, 6.119±.719 ; Area II, 5.475±.272 ; Area III, 5.191±.286 ; Area IV, 5.750±.184. Sample mean lie within +.445, +.543, +.555, and +1.687 (population) standard deviations from the respective population means. Except for the difference between means in Area IV—the fastest growing area between 1961 and 1971, these sample means are reasonably accurate estimators of the population parameters.

survey, is 588.56 rupees per month, or about $ 73.50 (Table 8.5). Rental, however, does not appear to consume a very large portion of incomes, perhaps no more than five per cent as a city-wide average. Male and female years of education average about eight and four, respectively. At the natural area level, the ecological correlations among these variables appear to be positive and moderately strong. Rental plus female education, as one pair, and income plus male education, as another, covary quite closely (this is borne out at the ward level also—data not reported here).

TABLE 8 5 : **Social rank variable (mean and standard deviation) for Ahmedabad and natural areas**

	Rent Rs. per month		Total household income (Rs.)		Years of education			
					Male Head		Female Head	
	Mean	S.D.	Mean	S.D.	Mean	S.D.	Mean	S.D.
Area I	30.43	30.12	583.54	147.51	7.49	1.98	4.59	1.71
Area II	29.78	24.18	527.50	96.73	7.42	2.76	3.76	1.92
Area III	22.43	15.22	554.71	113.17	7.87	2.62	3.44	1.98
Area IV	32.85	7.00	909.00	340.83	10.41	1.50	6.77	1.99
City (Mean)	28.49		588.56		7.77		4.27	

The most frequently-reported occupations are in the skilled-formal (blue-collar) and lower clerical and professional-formal sectors, respectively (Table 8.6). Just over seven per cent of the reported occupations were classified as professional and managerial, while just over eight and one-half per cent are in lower class categories (I and II). At the natural area level there are strong positive correlations between percentages in lower class and blue collar occupations and percentage unemployed ; and negative correlations between these and percentage in professional and managerial occupations. The ecological correlations between occupational structure and the social rank variables strongly indicate the association between high income, education, rental, high percentages in professional and managerial occupations and low percentages in lower class and blue collar occupations.

Table 8.7 lists information about family structure/urbanization variables for Ahmedabad and the four natural areas. We

TABLE 8.6 : **Percentage of male head of households by occupational categories (mean and standard deviation) for Ahmedabad and natural areas**

	Unskilled informal sector		Semi-skilled formal sector		Skilled formal sector		Lower clerical, professional formal sector	
	Mean	S.D.	Mean	S.D.	Mean	S.D.	Mean	S.D.
1	2	3	4	5	6	7	8	9
Area I	3.20	5.67	4.75	3.89	29.98	16.97	28.42	18.48
Area II	4.68	1.07	4.73	1.46	52.55	14.99	15.18	2.64
Area III	3.77	4.04	4.41	2.85	51.77	21.01	20.81	10.04
Area IV	2.80	.57	10.30	3.40	9.10	0.51	39.70	15.70
City (Mean)	3.50		5.08		37.97		24.59	

Managerial and professional		Retired		Unemployed		Not classifiable	
Mean	S.D.	Mean	S.D.	Mean	S.D.	Mean	S.D.
10	11	12	13	14	15	16	17
6.53	6.31	10.45	7.20	1.48	2.24	10.30	6.42
6.00	7.71	2.58	5.86	2.78	1.72	5.68	5.49
6.20	7.77	4.96	2.94	2.74	1.75	3.86	4.40
19.85	23.54	4.80	0.00	1.20	.57	9.15	5.02
7.33		7.06		2.02		7.46	

TABLE 8.7 : **Family structure characteristics (mean and standard deviation) for Ahmedabad and natural areas**

	Age Male Head		Persons per household		Percentage born		
	Mean	S.D.	Mean	S.D.	Ahmedabad	Gujarat village	Other
Area I	49.7	4.1	6.12	.72	63.06	11.43	25.51
Area II	45.7	3.3	5.48	.27	27.90	40.85	31.25
Area III	42.7	5.0	5.19	.29	35.21	29.70	35.18
Area IV	41.8	1.2	5.75	.18	37.30	21.00	41.70
City (Mean)	46.7		5.79		47.15	22.73	30.12

note that for the city as a whole and for all of the areas except the walled city, over half of the male heads of households were born outside of Ahmedabad, and about 23 per cent are village-born. The average age of the male heads is about 47 years, with the oldest population living in the walled city and the youngest on the west bank. The average number of persons per household reported in this survey is 5.79, slightly above the 1971 census average. A weak positive ecological correlation exists between age and percentage born in Ahmedabad, while a curvilinear (U-shaped) correlation appears to hold between these variables and persons per household. Percentage born in villages and persons per household are negatively correlated (suggesting that, not withstanding differences in age structure, large households is an *urban* trait in Ahmedabad).

Table 8.8 lists the percentages reported in the five most frequent responses to the question "what is your caste (i.e., *jathi*) ?" The survey indicates that about 14 per cent of the ward populations of Ahmedabad is Patel (including several sub-castes). Brahmins constitute about 10 per cent of the population, while Parmars—a scheduled caste—make up about 4 per cent. The Muslim households in the sample constitute about 6 per cent of the total, and an overwhelmingly high percentage of these households is in the walled city. A fairly strong ecological correlation exists between the presence of Brahmin and Baniya households, while percentage Brahmins and percentage Parmars are negatively correlated.

TABLE 8.8 : **Percentages of households in most frequently reported "Castes" (mean and standard deviation) for Ahmedabad and natural areas**

	Patel		Brahmin		Baniya		Muslim		Parmar	
	Mean	S.D.	Mean	S.D.	Mean	S.D.	Mean	S.D.	Mean	S.D.
Area I	11.0	12.9	10.9	8.7	9.7	11.2	12.4	14.0	1.3	1.9
Area II	9.2	4.8	9.9	7.7	7.6	4.3	0.2	.4	11.4	2.9
Area III	28.0	18.9	8.3	8.2	4.2	5.9	1.3	1.8	4.8	5.6
Area IV	9.5	13.4	11.9	7.9	15.9	2.3	—	—	0.4	0.6
City (Mean)	14.3		10.2		8.5		6.3		4.3	

In the following section the variation in the census- and survey-based social structural characteristics of Ahmedabad, in relation to the four natural areas are examined. The history of the city and the evolution of the walled city and the peripheral areas does appear to have a counterpart in the information examined already. Next, some of the details of the geographic distribution of Ahmedabad's contemporary social structure are described.

THE SOCIAL COMPOSITION OF AHMEDABAD'S NATURAL AREAS

Census data

The analysis of 1961 and 1971 census data (Tables 8.1, 8.3 and 8.4) suggests that Ahmedabad's four broad natural areas do correspond to social areas. This social area formation further indicates the characteristic interpenetration of modern and traditional elements in Ahmedabad's social structure. Not only do differences (some, but not all, statistically significant) exist between the values of natural areas on census variables, but the ordering of these differences is remarkably consistent.

The walled city has the highest ward-wise means for the variables persons per household, general literacy, female literacy, and percentage employed in trade and commerce. This area has the lowest ward means for percentage scheduled tribes, percentage scheduled castes, percentage employed, and percentage employed in manufacturing. It is, thus, an area which is high in social rank composition by modern (i.e., class) and traditional (i.e., caste) standards. Its households tend to be large (about 0.4 person per household larger than the mean for the city in 1971), to have few members in the labour force, and to have a high proportion of literate members (including females). While the area- and division-wise means on these variables do conceal considerable internal heterogeneity, they also reflect the historical continuity of the walled city as a settlement dominated by an educated, upper-caste commercial elite.

Area II, the near-peripheral area, contrasts sharply with the walled city. It has the lowest ward-wise means for general literacy and female literacy and the highest means for percentage scheduled castes, per cent working population, and per cent working females. This area is also next to lowest in per cent employed in trade and commerce, in other services and in persons per household ; and is next to highest in percentage employed in agriculture, in manufacturing and percentage scheduled tribes. The mill area is thus an area of low social rank in which a high proportion of males and females is employed. Although this area has a large blue collar segment, the majority of the population is employed in non-industrial (and non-commercial) pursuits. By modern standards, it is a lower (as opposed to working) class area in which literacy rates are very low (about 13 per cent below the general rate for the city). By traditional standards, it is a lower caste and rural-oriented neighbourhood. The fact that all of the wards in this area are bounded by the wall suggests that the traditional exclusion of socially inferior groups from the old city persists, at least at an aggregate level.

Area III, the far peripheral area, is similar to the mill section in many respects, although its social rank and related characteristics appear to be slightly, but consistently, higher. This similarity is, in part, due to the artificiality of the physical boundary between the two peripheral areas. Much of the inner portion of the wards in Area III might well be classified as mill area, whereas the outer portion of these wards is dominated by middle class housing and the other features mentioned in earlier sections, ward and block level analysis also suggests that Area III consist of two or more relatively distinct sub-social areas ; nevertheless area-wise means do reflect an interesting contrast between the walled city and the newer sections. Area III has the lowest mean number of persons per household, the lowest percentage employed in trade and commerce and in other services, and the highest percentage employed in manufacturing. It is also next to highest in percentage scheduled caste and next to lowest in female and general literacy. It is slightly above the city mean in per cent working population, but slightly below the city mean in per cent working females. This combination of characteristics suggests that, at the aggregate level, Area III is the principal working class neighbourhood of Ahmedabad. It is a moderately low ranking area in relation to both modern and traditional characteristics, although it appears to be more modern than the near peripheral area (i.e., more blue collar workers, higher literacy rates, a smaller percentage of scheduled castes).

Census data for Area IV, the west bank, reflect both the area's rural heritage and its current role as Ahmedabad's suburbia. It is next in ranking to Area I, the walled city, on many of the variables analyzed : persons per household (high), female and general literacy (high), per cent employed in trade and commerce (high), per cent scheduled caste (low), per cent working population (low) and per cent working females (high). It is the lowest of the four natural areas in percentage employed in manufacturing, and the highest in percentage Scheduled Tribe and percentage employed in agriculture. Its ratio of percentage employed in trade and commerce to percentage employed in manufacturing of 0.90 is the highest of the four areas (city ratio =0.50, ratio for Area III=0.27). Thus we see that the west bank is a middle class neighbourhood which, though in some ways lower in social rank than the walled city, has a large modern commercial-employed sector. At the same time, it is the most rural of the four areas, not only in terms of its geography and history ; but also in relation to its occupational and ethnic (i.e., tribal) composition. Thus, while this area is one in which we are very likely to find suburban, white collar households, it is also one in which we will find relatively many households whose social affiliations and economic activity are indistinct from those of the peasants in the surrounding countryside.

A general pattern appears to exist among the four natural areas in relation to ward-wise values on census variables. Area I, the walled city, is highest in social rank, literacy, household size and caste composition. Next highest in these respects is Area IV, the west bank. Yet, Area IV is also the most rural while Area I the least rural of the four. Ranking below Area IV in social rank and related variables is Area III, the far peripheral zone, and last in these respects are the wards which border on the wall. This pattern, I-IV-III-II, does form a rough concentric zone configuration in space, but perhaps of greater importance is the correspondence between social and natural area formations in Ahmedabad's history. The relationship between the wall, the mills, the west bank, etc. reflect the copresence of modern and traditional social geographic principles in a most vivid sense—at least to the limits of the validity of the census variables.

In order to summarize some of these observations, factor scores were derived from the analysis of 1961 census data reported in Table 8.3. Table 8.9 shows the ward-wise means and standard deviations of factor scores for the four netural areas (since the scores are standardized, the city mean on all four of these factors is 0.0). We see that Area IV, the west bank, is clearly the highest, followed by area II, then I and III. A high score on this factor (orthogonally rotated) indicates a high percentage of working females, combined with a high percentage working in agriculture and extraction, moderately high percentages in scheduled castes and tribes and moderately low literacy rates. This is a complex combination of variables and ecological correlations, a fact which is reflected in the lack of correspondence between the geographic distribution of factor I and of the single social rank, ethnicity and family structure characteristics. Since the first factor in 1971 is somewhat less

TABLE 8.9 : **Factor scores by natural area means and standard deviations (1961)**

Natural Area	Factor I Mean	Factor I S.D.	Factor II Mean	Factor II S.D	Factor III Mean	Factor III S.D.	Factor IV Mean	Factor IV S.D.
I	—.4514	.713	.3598	.638	.5842	.344	—.1917	.255
II	.8121	1.122	.0018	1.346	—1.0646	1.256	—.2903	.755
III	—.6135	.538	—.6731	.939	—.5093	.894	.0660	.640
IV	1.8476	.010	.3482	1.615	.9337	.855	1.8527	3.393

ambiguous, we would expect its geographic distribution to more closely approximate those of the single variables (although this analysis is not reported here).

Factor II, which has been characterized as an urban white collar/blue collar dimension, displays a more characteristic I-IV-II-III pattern. Both the walled city and the west bank are above the city mean on this factor, and their ward-wise means are nearly identical. At the other extreme, Area III, with its low white collar/blue collar ratio averages more than two-thirds of a standard deviation below the city mean on this factor. Area II is at the city mean, although its relatively low blue collar concentration places it above Area III.

Factor III, a combined white collar-urbanization index, also displays a variant of the characteristic geographic patterning. In this instance, Area IV has the highest average factor scores (nearly 1.00 above the mean), followed by Area I, then III and II. The high rank of the west bank on this factor is a function of the low percentage of working population and the high percentage of workers in trade and commerce in the area. These elements, in addition to high loadings on persons per household, and general and female literacy, also explain the high score for the walled city and the low scores for the peripheral areas. Factor IV, a transport workers factor, displays a IV-III-I-II pattern.

The geographic distributions of census factor scores and, especially, of single census variables suggest particular types of social area formation among Ahmedabad's natural areas. In order to examine this correspondence between social and natural areas in a more rigorous manner, two significance tests were performed on the differences between (ward-wise) means and variances for the natural areas. Tables 8.11 and 8.13 report the results of *t*-tests (difference between means) using 1961 and 1971 census data, respectively. In these tables are listed the variables and *p* values for which the probability is less than .05 that the observed difference between means for two natural areas is due to chance alone. Tables 8.12 and 8.14 report the results of *F*-tests (analysis of variance) for 1961 and 1971. These tables list the variables and *p* values for which the probability is less than .05 that the observed *F*-ratio (of between/within variance) for two natural areas is due to chance alone. Together, these tables identify the variables which are the true discriminators between natural areas.

Table 8.10 indicates that for 1961 the greatest number of significant differences exist between the walled city and the mill area which borders it ; the second greatest contrast is between the walled city and the far peripheral zone. In this respect, Area I's distinctness as a predominantly upper-middle class, upper caste and highly literate section of the city is verified. Supportive of this is the relative lack of difference between Area I and Area IV. These two areas differ significantly only in relation to percentage scheduled caste (Area I has the lower percentage, 3.7 vs. 9.0) and two of the factors, I and II : differences which reflect the markedly rural, non-industrial character of Area IV. The few differences which exist between Areas II and III underscore the higher social rank, in class and caste senses, of Area III. In

TABLE 8.10 : **t-test significant (.05, 2-tail) differences between natural areas, 1961 census variables and factor scores**

Variables	P
Natural Areas I/II	
Persons per household	<.001
Percentage scheduled caste	.024
Percentage literate females	<.001
Percentage working females	.040
Percentage working population	.049
Percentage working in trade and commerce	.009
Percentage literate population	<.001
Factor I	.008
Factor III	<.001
Natural Areas I/III	
Percentage literate females	.010
Percentage working population	.004
Percentage working in manufacture	.002
Percentage working in trade and commerce	<.001
Percentage literate population	.016
Factor II	.007
Factor III	.004
Factor IV	.006
Natural Areas I/IV	
Percentage scheduled caste	.003
Factor I	.001
Factor IV	.016
Natural Areas II/III	
Percentage scheduled caste	.043
Percentage working females	.026
Factor I	.008
Natural Areas II/IV	
Percentage literate females	.027
Natural Areas III/IV	
Factor I	<.001

general, the I-IV-III-II ranking of Ahmedabad's social areas appears to be substantiated by these *t*-tests.

Table 8.11 indicates that for 1961 the differences observed between Area I and all of the other areas significantly exceed within variances on the greatest number of variables. Thus, the walled city appears to be not only a distinct, high-ranking, neighbourhood, but it is also the most homogeneous neighbour-

TABLE 8.11 : **F-test significant (.05, 2-tail) differences between natural areas, 1961 census**

Variables	P
Natural Areas I/II	
Percentage scheduled caste	<.001
Percentage working population	<.001
Percentage working in agriculture	<.001
Percentage working in manufacture	.028
Factor II	.032
Factor III	<.001
Factor IV	.002
Natural Areas I/III	
Percentage scheduled caste	<.001
Percentage literate females	.020
Percentage working population	.005
Percentage working in transport	.001
Factor III	.004
Factor IV	.006
Natural Areas I/IV	
Percentage working population	<.001
Percentage working in agriculture	.001
Percentage working in manufacture	.003
Percentage working in transport	<.001
Factor I	.023
Factor IV	<.001
Natural Areas II/III	
Persons per household	<.001
Percentage scheduled tribe	.019
Percentage working population	.034
Percentage employed in agriculture	<.001
Natural Areas II/IV	
Percentage working in transport	<.001
Factor I	.014
Factor IV	.013
Natural Areas III/IV	
Percentage working population	.009
Percentage working in agriculture	.004
Percentage working in manufacture	.033
Percentage working in transport	.001
Factor I	.030
Factor IV	.002

hood (at the ward level). The variables which yielded significant *F*-ratios for the walled city (note the low *P*-values) show that its upper-caste, non-manufacturing, middle class character is consistent throughout the area. Area IV, the west bank, appears to be the next most homogeneous area (although the fact that it consists of only two wards make these significant *F*-values somewhat doubtful indicators of homogeneity). Areas II and III appear to be the most similar, although they do differ significantly on variables related to urbanization and modernity.

The 1971 census *t* and *F*-tests reported on Tables 8.12 and 8.13 indicate that the significant differences observed in 1961 have, in general, persisted. Again the high rank of Areas I and IV and the low rank of the two peripheral areas are verified. The low

TABLE 8.12 : t-test significant (.05, 2-tail) differences between natural areas, 1971 census

Variables	P
Natural Areas I/II	
Persons per household	.022
Percentage scheduled caste	.005
Percentage literate females	.001
Percentage working population	.009
Percentage working in trade and commerce	.045
Percentage literate population	<.001
Natural Areas I/III	
Persons per household	<.001
Percentage scheduled caste	.013
Percentage literate females	.020
Percentage working females	.001
Percentage working in agriculture	.004
Percentage working in manufacture	<.001
Percentage literate population	.021
Natural Areas I/IV	
Percentage scheduled caste	<.001
Percentage working in manufacture	.013
Natural Areas II/III	
Percentage scheduled caste	.015
Percentage working females	.009
Natural Areas II/IV	
Percentage scheduled caste	.014
Percentage literate females	.029
Percentage working in agriculture	.007
Natural Areas III/IV	
Percentage working in agriculture	.020
Percentage working in manufacture	.051
Percentage working in trade and commerce	.051

TABLE 8.13 : **F-test significant (.05, 2-tail) differences between natural areas, 1971 census**

Variables	P
Natural Areas I/II	
Percentage scheduled caste	<.001
Percentage working population	<.001
Percentage working in agriculture	<.001
Percentage working in manufacture	.006
Percentage working in transport	<.001
Natural Areas I/III	
Percentage literate females	.009
Percentage working in agriculture	.008
Percentage working in manufacture	.001
Percentage working in transport	.022
Natural Areas I/IV	
Percentage working in transport	<.001
Natural Areas II/III	
Percentage scheduled caste	.034
Percentage scheduled tribe	.024
Percentage working in manufacture	.042
Percentage working in transport	.027
Natural Areas II/IV	
Percentage scheduled caste	.023
Natural Areas III/IV	
Percentage working in transport	.005

percentages of scheduled castes, workers in manufacturing and percentages employed and the high percentages literate and employed in trade and commerce in the walled city continue to be significant. Areas I and IV continue to be similar in social rank, yet the more commercial and rural character of the west bank persists. The *F*-tests indicate that these two areas are becoming less internally homogeneous in relation to each other, as the percentages of the populations employed, working in agriculture and working in manufacturing have become more evenly distributed at the ward levels (percentage employed and percentage working in manufacturing both decreased between 1961 and 1971, but percentage working in agriculture increased). No major changes appear to have occurred between 1961 and 1971 in the relationship between Areas II and III, whereas the differences between these areas and the other two did persist.

Survey data

The correspondence between social and natural areas observed in the geographic distribution of census variables can also be seen in the distribution of survey variables, as reported in Tables 8.6—8.9. Because the survey was designed to provide relatively unambiguous information about social rank, family structure (urbanization) and ethnicity, additional information about Ahmedabad's social area formation is accessible. Some of the differences between the walled city and the west bank which were unclear from an examination of census data do become more apparent with the use of survey data ; and other details of the city's social geography are revealed in this way.

The generally high social rank of Areas I and IV is basically verified in the class-like occupational distributions, the incomes, rentals and levels of educational attainment of these areas. Overall, the west bank does appear to have the higher rank of the two, with an average total household income of over 900 rupees per month and an average of more than ten years of education (eleven years is equivalent to high school graduation) for male heads. Rentals, years of education for female-heads, and the percentages of male heads in the two middle-class occupational categories (lower clerical and professional and professional and managerial) for Area IV are also the highest in the city by substantial amounts. The walled city is next highest in rentals, total household income, years of education for female heads, and percentages in middle class occupations. Area IV has the lowest percentage in the three lower class categories—only about nine per cent blue collar—and Area I has the next lowest, with about 30 per cent blue collar. The only serious discrepancy between survey and census findings on social rank is the relatively low number of years of education for male heads in the walled city of about 7.5 (although the male education means in all three east bank natural areas are close to the city mean of 7.77 years).

The lowest mean incomes, years of education (male) and percentages in middle class occupations are found in Area II, while the lowest rentals and years of education (female) are in Area III. Percentages in the lower class occupations in the two peripheral areas are virtually the same (with slightly more than 50 per cent of these areas employed in blue-collar occupations).

Once again, it confirms the relatively low rank and the relative social homogeneity of these two areas ; while the contrast between Areas II and IV is especially sharp.

Family structure data from the survey, that is, ages, birth places and persons per household, are extremely consistent with census findings. The characteristic I-IV-III-II pattern of rank-ordering is evident with respect to ages of male heads of households (the walled city with the oldest population), persons per household (the walled city with the largest number), and percentages born in Ahmedabad (the walled city with the highest percentage). This pattern is altered to IV-I-III-II in the case of percentages of male heads not born in Gujarat villages (i.e., the west bank had the highest percentage village-born), just as it appeared to be with the census variables indicating degree of ruralness. Thus, Area I can be seen as a neighbourhood with an older, native-born urban population and relatively large households (partly because of higher fertility, partly because of higher proportions of extended-family households).

An examination of the distribution of Ahmedabad's five most frequently reported *jathi* shows that intra-areal variances are generally high. This indicates that while caste segregation does exist, it tends to occur at ward and even block levels, rather than by natural area. In this sense, each natural area has a distinct system of *jathi* which is, to some extent, repeated in the other areas. The principal exception to this in the case of Ahmedabad is the Muslim population, which appears to be highly concentrated in the walled city and very thinly distributed or virtually absent in the other areas. There surely is an historical continuity reflected in this, as the Muslims who are the original dominant ethnic group of Ahmedabad have yet to settle outside of the city in any significant numbers.

While the Brahmin population is rather evenly distributed throughout the wards and natural areas of Ahmedabad, the highest concentrations are found in Areas IV and I (respectively), while the lowest is in Area III. The lowest concentrations of Parmars are in Areas IV (0.4%) and I (1.3%), respectively ; while the highest is in Area II, with nearly three times the percentage of the city as a whole. The Baniya population is relatively highly concentrated in Area IV, Area II has the next highest percentage of Baniyas while Area III has the lowest percentage.

The general pattern of caste distribution between Ahmedabad's natural areas indicates an upper caste Hindu orientation in the west bank wards with a weaker but still upper caste orientation in the walled city. The most frequently reported group in the walled city is Muslim, while the percentage of Patels in the walled city also exceeds that of the west bank. The absence of Muslims in the other natural areas and the high concentration of Patels in Area III further underscore the distinctiveness of Area I. Areas IV, II and I (in order) reflect domination of Hindu ranking systems ; Area I, while still predominantly Hindu (and upper rank) has a larger, traditional non-Hindu component.

Tables 8.14 and 8.15 report the results of *t* and *F*-tests performed with survey data. The *t*-tests indicate that Area I differs significantly from the other areas with respect to both family structure/urbanization and caste characteristics. The difference in percentages in blue collar occupations between Area I and Area III is also significant. While Area I and IV do differ in age, birth places and percentage Muslims in the observed manner, the mean incomes in Area IV point to a significantly higher social rank. Area IV differs significantly from the other areas principally with respect to social rank, especially occupational characteristics. There are no significant differences reported, for these variables, between Areas II and III.

The *F*-tests indicate that the walled city and the two peripheral areas are distinct with respect to place of birth, percentages employed in both middle class and one of the lower class occupations, and percentage Muslim. The only variable with a significant *F*-ratio for Area I vs. Area IV is percentage employed in managerial and professional occupations, indicating again an increasing similarity between these two upper-ranking neighbourhoods (Principally because of the increase in rank of the west bank). Areas II and III appear to be significantly distinct with respect to two caste characteristics.

Overall, the significance tests reflect the patterns observed in the analysis of census data and the overview of the survey variables. Area I is highly urban, high in social rank in traditional respects and distinctly Muslim. Area IV is very high in modern social rank characteristics, distinctly Hindu and somewhat rural. Both areas have relatively few Parmars, relatively many Brahmins and high proportions in middle-class occupations, although there

TABLE 8.14 : **t-test significant differences between natural areas ; selected survey variables**

Variables	P
Natural Areas I/II	
Percentage male heads born in Ahmedabad	<.001
Percentage male heads born in Gujarat villages	<.001
Percentage employed in skilled manual occupation	.027
Percentage Parmar (scheduled caste)	<.001
Percentage Muslim	<.001
Natural Areas I/III	
Age of male head	.003
Percentage male heads born in Ahmedabad	<.001
Percentage male heads born in Gujarat villages	.002
Percentage employed in skilled manual occupation	.018
Persons per household	.004
Percentage Patel	.027
Percentage Muslim	<.001
Natural Areas I/IV	
Age of male head	.023
Percentage male heads born in Ahmedabad	.011
Percentage male heads born in Gujarat villages	.042
Total household income	.026
Percentage Muslim	<.001
Natural Areas II/IV	
Percentage male heads employed in semi-skilled occupation	.038
Percentage male heads employed in lower clerical, professional occupation	.026
Percentage Parmar	.007
Natural Areas III/IV	
Percentage males employed in semi-skilled occupation	.041
Percentage males employed in skilled manual occupation	.029
Total household income	.032
Persons per household	.038
Percentage Baniya	.034

is a large blue collar component in the walled city. Areas II and III are fairly similar in social rank and family structure characteristics, although there is a moderate tendency toward caste segregation between the areas. Both of the peripheral areas are predominantly Hindu, but with a lower caste skew. A very high proportion of Ahmedabad's Parmars are found in Area II.

TABLE 8.15 : **F-test significant differences between natural areas ; selected survey variables**

Variables	P
Natural Areas I/II	
Percentage male heads born in Gajarat villages	.007
Percentage males in unskilled, informal occupation	.019
Percentage males in lower clerical, professional occupation	.008
Percentage Muslim	<.001
Natural Areas I/III	
Percentage male heads born in Gujarat villages	.002
Percentage males employed in professional and managerial occupation	.038
Persons per household	.033
Percentage Parmar	.002
Percentage Muslim	<.001
Natural Areas I/IV	
Percentage males in professional and managerial occupation	.005
Natural Areas II/III	
Percentage Patel	.046
Percentages Muslim	.020
Natural Areas II/IV	
Percentage males in lower clerical, professional occupation	.019
Natural Areas III/IV	
Percentage males in skilled manual occupation	.041
Percentage males in professional and managerial occupation	.046
Total household income	.047

CONCLUSIONS

This examination of Ahmedabad's social structure and geographic distributions has revealed three general features of the city. First, there is a persistent co-presence of traditional and modern, urban and rural, upper-rank and lower-rank characteristics. Second, there is a lack of a clear tendency toward modernization in the recent social changes for which census and survey data are available. Third, there is a close relationship between the historical and geographic characteristics of Ahmedabad's natural areas and their present social composition. These results indicate that the traditional/local and modern/western elements which have contributed to the evolution of Ahmedabad are present and active in a city's contemporary ecological structure. It is a city which has grown by accretion and accommodation, rather than by replacement. This appears to be true in physical and, based upon these findings, even in social senses.

When Louis Wirth and the other Chicago School sociologists characterized urban life in terms of its density, complexity and heterogeneity, their principal cases were American and European. The development of models and theories of urban ecological structure based upon the work of the Chicago School extended these observations in providing an increasingly organized (and a more empirically grounded) sense of these characteristics. In a city like Ahmedabad, which in addition to the familiar sources of density, complexity and heterogeneity, has also been influenced by colonial and imperial relations, conquest, and conscious modernization policies, the classic models and theories prove to be excessively simple (or inadequate in other ways). The order of density, complexity and heterogeneity of Ahmedabad far surpasses that anticipated by existing models. In some respects, it is the most profoundly urban of any of India's large cities.

In attempting to discern the patterns and regularities in Ahmedabad's ecological structure, it was found useful to begin with very broad, and in some senses very crude, natural area formations. Yet, despite the generality of the findings reported here, the physical features of Ahmedabad such as the wall, the mills and the river have proved to have both traditional and contemporary social impacts. These preliminary results suggest, as Joseph Gusfield (1967) observed some time ago, that tradition and modernity may be false polarities in the study of social change. For :

> 'Tradition' and 'modernity' are not disjoint except in ideal typical discussions. They manifest themselves together . . . as perhaps they always have (Ramu and Wiebe 1970, p. 138).

This is especially clear when, in cases like Ahmedabad, traditional and modern cultures visibly co-exist in urban architecture and geography.

by replacement. This appears to be the principal one, based upon these findings, even in social science.

While Louis Wirth and the other Chicago School sociologists characterized urbanism in terms of its density, complexity and heterogeneity, their principal cases were American and European. The development of models and theories of urban ecological structure based upon the work of the Chicago School exerted these city studies in providing an increasingly empirical and a more empirically grounded sense of these characteristics. In a city like Ahmedabad, which in addition to the familiar sources of density, complexity and heterogeneity, has also been influenced by colonial and imperial relations, contacts and conscious modernization policies, the classic models and theories prove to be excessively simple (or inadequate in other ways). The order of density, complexity and heterogeneity of Ahmedabad far surpasses that anticipated by existing models. In some respects, it is the most profoundly urban of any of India's large cities.

In attempting to discern the patterns and regularities in Ahmedabad's ecological structure, it was found useful to begin with very broad, and in some sense very crude, macro-level formulations. Yet, despite the generality of the analyses reported here the physical features of Ahmedabad such as the walls, the mills and the river, have proved to have both traditional and contemporary social impacts. These preliminary results suggest, as Joseph Gusfield (1967) observed some time ago, that tradition and modernity may be false polarities in the study of social change. For:

> 'Tradition' and 'modernity' are not distinct except in ideal typical discussions. They manifest themselves together ... as perhaps they always have (Ra[illegible] and We[illegible] 197[illegible], p. [illegible]).

This is especially clear when, in cities like Ahmedabad, traditional and modern features clearly coexist in urban structures and processes.

NINE

V.L.S. PRAKASA RAO AND VINOD K. TEWARI

THE ECOLOGICAL STRUCTURE OF BANGALORE CITY

THE ecological structure of a city is a product of social, economic and locational characteristics of its inhabitants and their activities, evolved over a period of time. Households of similar socio-economic status, and similar activities tend to cluster at various locations within the city. Some elements tend to cluster at different distances from the city centre, while the others along the transport corridors, or, at 'choice locations'. The end result is characterised by socio-economic isomorphism and spatial stratification/segregation.

The structure of many Indian cities has evolved as a result of both indigenous as well as colonial traditions. This has given rise to a dualism in the city structure which is best illustrated in the highly mixed land use, sharp density ratios between older areas and the periphery, co-existence of 'Firm-type capitalist economy' and 'Bazaar-peasant economy', and differences in life styles and social organisation of the population (Breese 1966, Smailes 1969, McGee 1971).

The highly mixed land uses in Indian cities, that differ from the segregated pattern of land uses in Anglo-American cities, is partly due to the necessity for a large proportion of population to walk between the place of residence and the places of work, shopping and services. Residential segregation in Indian cities is also not as clearly identifiable as in the Western cities. Only the population groups at the polar ends of the social status hierarchy can be found living in segregated homogeneous areas.

All the other groups often live in mixed status neighbourhoods, and yet, remain socially segregated by adhering to strict social codes. In the households' choice of residential location, social factors like religion, caste and language usually get higher weightage than economic factors (occupation, income, etc.). Very often, houses in a residential area are available at a lower rent than the prevailing market rent, if the household desiring the accommodation belongs to a particular caste, religion, language or regional group. Sometimes, even the food habits (like, being a vegetarian) are given additional consideration. As a result, even if a particular socio-economic group predominates in a residential nucleus, all income levels are represented there. Thus, the structure that emerges out of such a complex interplay of physical, social, economic, cultural, and behavioural factors, does not follow the classical "rich centre and poor periphery" type of models.

To examine to what extent does the pattern of the ecological structure of Bangalore follow the classical model, the following analysis is based on the structure in the historical past of the city, and spatial distributions of selected socio-economic characteristics. This study based on a multi-diciplinary household survey (sample size : 1745 households) of Bangalore city conducted by the authors in 1974, draws heavily from the published report of the survey (Prakasa Rao and Tewari 1979).

SOCIO-ECONOMIC PROFILE OF THE CITY

Bangalore City* like any other Indian city retains its rural traditions. The traditional elements can be traced to religion, language, and caste based socio-economic spatial segregation within the city in spite of the dominance of secondary and tertiary sectors.

In terms of social composition, the city's cosmopolitan character gets reflected in its socio-cultural diversity with six religions : Hindu (81%), Muslim (19%), Christian (7%), Jain, Sikh and Zorastrians (together 2%) ; 217 castes/sub-castes among

* Bangalore, the capital city of Karnataka State, with its population of 2.9 million (according to 1981 census) ranks fifth in the city system of India.

Hindus with Agricultural Castes dominating (24%), followed by Brahmins (21%), Artisans and servicing castes (17%) and scheduled castes (17%) ; and 17 languages with Kannada (34%) as the dominant language followed by Tamil (25%), Telugu (17%), and Urdu (19%). Though all religion and caste groups showed tendency to move away from traditional occupations and lower levels of education, yet there is polarization at the upper levels in educational and occupational status. Thus overall social distance persists, with a slow upward mobility.

According to the Bangalore Survey, the city mainly consisted of the middle class (household income Rs. 300-999) households (59%), though 24 per cent of the total households lived with a monthly income of less than Rs. 300, 13 per cent with Rs. 1,000 to Rs. 1,999, and 4 per cent with Rs. 2,000 and above (Fig. 9.1). The mean income for the whole sample was Rs. 657 per month, and ranged between Rs. 11 in the lowest income class and Rs. 2,902 in the highest income class (Table 9.1 and Fig. 9.1). The

TABLE 9.1 : **Percentage distribution of households and their incomes by monthly household income class**

Household income (Rs.)	Percentage of households	Mean income (Rs.)	Cumulative percentage of households	Cumulative percentage of total income of the households
50	0.3	11	0.3	0.0
50— 149	2.2	110	2.5	0.4
150— 299	21.8	224	24.3	7.8
300— 499	27.0	374	51.3	23.1
500— 749	22.3	598	73.6	43.4
750— 999	9.2	853	82.8	55.3
1,000—1,999	12.9	1,306	95.7	80.9
2,000+	4.3	2,902	100.0	100.0

total income was unequally distributed (concentration ratio : 0.406).

Religionwise, mean income of households belonging to Muslim religion was the least (Rs. 540 per month). It was Rs. 652 for Hindus and Rs. 702 for Christians. The highest mean income

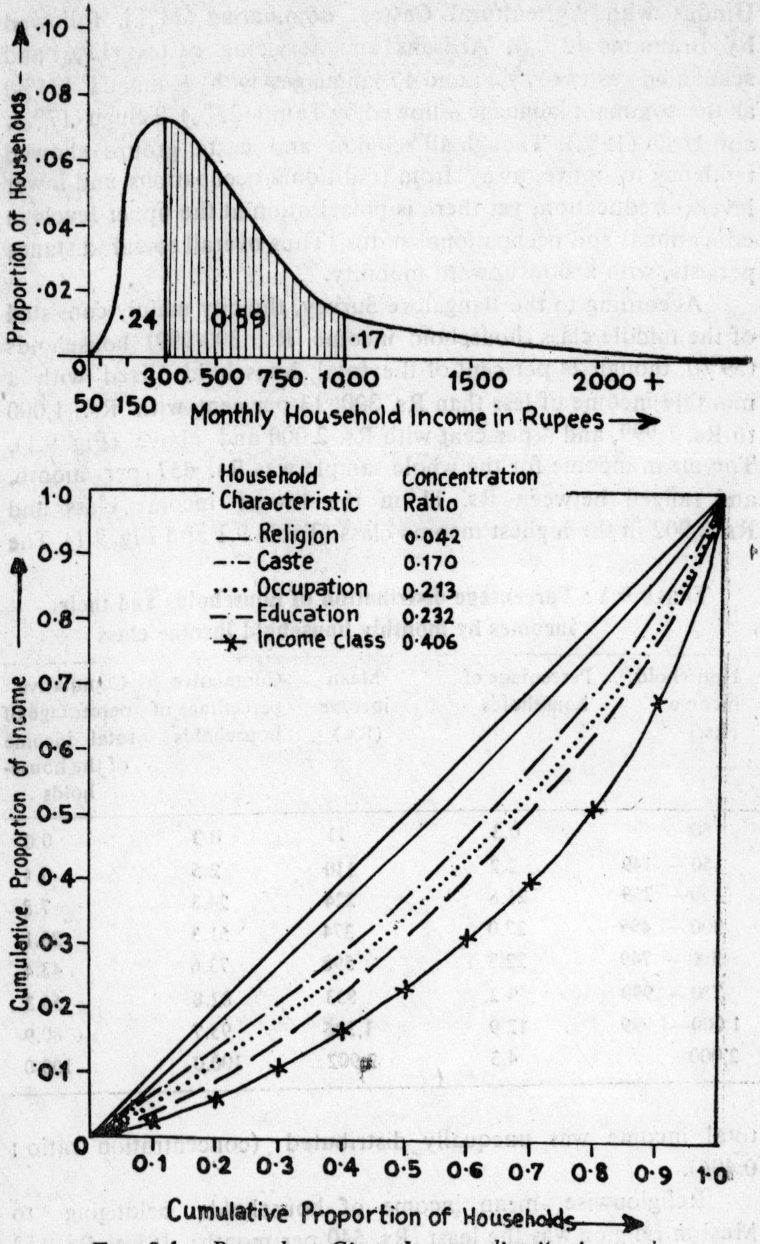

FIG. 9.1 : *Bangalore City—Income distribution and concentration*

(Rs. 1,504) was of the households belonging to other religions most of whom were Jains (Table 9.2). Inequality in the income distribution with respect to religion was least (concentration ratio : 0.042).

TABLE 9.2 : **Percentage distribution of households by selected socio-economic characteristics and monthly household income**

Socio-economic characteristic	Household income class				Mean income (Rs.)
	<300	300-999	1,000+	Total	
Religion					
Hindu	84.1	79.8	78.2	80.6	652
Muslim	10.0	11.6	7.4	10.5	540
Christian	5.2	7.4	10.4	7.4	702
Others	0.7	1.2	4.0	1.5	1,504
Total	100.0	100.0	100.0	100.0	657
Caste					
Brahmins	3.7	20.4	46.8	20.6	1,000
Lingayats and trading castes	13.2	15.6	18.5	15.5	752
Agricultural warrior and servicing castes	53.2	47.4	29.6	45.9	550
Scheduled castes	28.2	15.6	2.1	16.5	393
Others	1.7	1.0	3.0	1.5	823
Total (Hindus)	100.0	100.0	100.0	100.0	652
Occupation					
Professional and administrative	1.5	14.4	52.1	15.7	1,245
Clerical	7.2	18.6	10.2	13.4	604
Sales	6.9	13.0	15.1	11.2	732
Production workers	77.3	39.9	7.2	47.8	454
Others (non-workers)	7.1	14.1	15.4	11.9	692
Total	100.0	100.0	100.0	100.0	657
Education					
Illiterates	13.8	4.6	1.3	6.3	405
Up to middle	74.4	48.8	16.5	49.4	483
Secondary	11.1	33.9	34.2	28.4	747
Graduates and above	0.7	12.7	48.0	15.9	1,141
Total	100.0	100.0	100.0	100.0	657
Residential Status					
Resident	41.0	38.0	32.9	37.9	612
Migrant	58.9	62.0	67.1	62.1	685
Total	100.0	100.0	100.0	100.0	657

The average household income among Brahmins was the highest (Rs. 1,000) followed by trading castes and Lingayats (Rs. 752), agricultural warrior/soldiers and servicing castes (Rs. 550), and scheduled castes (Rs. 393). The inequality measure in the income distribution with respect to caste was 0.170.

Mean income varied with the occupation of the household head. It was the highest for the households with professional and administrative workers as head (Rs. 1,245).

URBAN SPREAD AND POPULATION DENSITY

The Fort near the City Market on the west, and the Barracks near the Ulsoor Tank on the east, were the two foundations of Bangalore City. The two kernels of the city, situated only 4 miles apart, were founded with a time-lag of more than two and a half centuries—the Fort built in 1537 and the British Garrison and Barracks established in 1809. The neighbourhood of the Fort and the neighbourhood of the Ulsoor Tank developed under two different *raisons d'etre*. The defensive strong point apart, the Fort city was oriented to the philosophy of a city and temple builder, an agriculturist turned warrior, under whose patronage developed a township of wholesale and retail traders, and of highly skilled artisans particularly families of cotton weavers (Hasan 1970, Rao 1930). In contrast to this, the neighbourhood of the Ulsoor Tank was developed to cater to the requirements of the British Troops and British officers who were pulled out of Srirangapatnam and settled near the tank. Thus, a military cantonment developed along with a civilian settlement of mainly retail traders and service classes adjoining the Ulsoor village. While the Fort neighbourhood developed as a typical native town with its bazaar, traders, and artisans, the Ulsoor neighbourhood developed as a cantonment with its artillery and cavalry barracks, parade ground, infantry road, the Mall, fine and spacious bungalows, bars and night clubs. In addition, polo-playing and racing developed in keeping with the "cantonment culture". Thus, the cantonment became the home for the British-Oriental Anglo-Indians (in the sense of Eurasians) and Indian Christians.

Both the nuclei expanded ; the Fort neighbourhood expanded east, north, and south, and the Ulsoor neighbourhood expanded west, north, and south, invading the open spaces and fertile agricultural lands. The east and west developed under two independent administrative units, with independent development programmes for nearly eight decades till their merger in 1949. Thus the development goals of both the municipalities were to provide minimal urban amenities and housing. The initial disparities in congestion, growth rates, and levels of urban amenities led to an east-west zonation.* With the development of the Russel Market and westward expansion of the cantonment, the nucleus of the west shifted to the Russel Market locality, while the development of the City Market near the Fort contributed to further intensification of the growing nucleus of the west. The east-west zonation certainly had its intermixing and integrating social-cultural elements in the two nuclei : temples, churches, and mosques near both the City Market and the Russel Market, though the east had more churches and mosques and the west more temples. Even the street names such as Dharmaraya Koil Street and Meenakshi Koil Street in the predominantly Muslim and Christian areas near the Russel Market, and Sultanpet near the City Market, a predominantly Hindu area, only echo the co-existence of the different religious communities.

From the initial pre-urban nucleus (the Fort) the earliest extensions were largely residential. Here, residential 'leap-frogging' was accompanied by residential infilling. The second phase of urban spread (1940 to 1960) was dominated by the 'leap-frogging' of large scale industrial complexes like Hindustan Aeronautics Limited (established in 1940)—a distance of about six miles to the east, Indian Telephone Industries (established in 1948)—about eight miles to the north-east, Hindustan Machine Tools (established in 1949)—almost seven miles north-west, and Bharat Electronics Limited (established in 1954)—about six miles north from the Fort. Infilling had taken place in the form of residential development in the interstitial areas, with the spread

* Although, the east-west differentiation in terms of economic characteristics is not so marked, it is quite significant in terms of social composition of the populations in the two parts of the city. According to the Bangalore Survey, while the west had higher percentage of Brahmins (almost 4 times that to the east), the east had a higher percentage of Christians, Muslims and Scheduled Castes (almost 2 times that to the west).

of the city, the urban texture lost its compactness. In 1956, Mysore State was reorganized with Bangalore as the capital of the reorganized State. This further opened the gates of the city, as it were, for the migrants both to the city and to its adjoining areas.

The next phase (1960 onwards) was one of "leap-frogging" of institutional and industrial complexes such as Bangalore University and the new extensions to industrial complexes, and the infilling of residential layouts, circumferentially to the city's built-up area to gain benefits of proximity to the city. One major consequence of the urban-spread from the beginning was the invasion into the green setting and open spaces, except into the defence areas. In 1889, the built-up area of the city was about 3.9 sq miles out of a total area of 18.1 sq miles. Thus, the area under parks and open spaces was more than four times the built up area. In 1974, this area was drastically reduced to 6.7 sq miles, with the city built-up area four times the area under open spaces.

In the process of urban spread (Fig. 9.2) of residential, commercial, industrial and institutional land uses, the main determinants were axial transport routes, particularly for industries, preference for locations near the outward moving edge of the city to avoid congestion, proximity to the places of work* and services, and the very availability of space in the interstitial areas and at the city edge. The one common denominator in the spread of the city was the highly mixed and haphazard growth of land uses in spite of the City Master Plan (Table 9.3). This can be attributed to the multiplicity of jurisdictions involved in the city development, and the uncoordinated and individual decisions of the private developers.

The gross population density for the city corporation area in 1971 was 37 persons per acre. It was as low as 12 per acre in 1901. While the city maintained its overall low gross density level over the decades, it did not escape from the imbalances between

* An analysis of movement for work in Bangalore City (Prakasa Rao and Tewari 1979) showed the following distribution of workers by mode of transport used in commuting from residence to work place : walk —46.4 per cent, cycle —12.4 per cent, bus —22.5 per cent, motorcycle/scooter —6.6 per cent, car—1.8 per cent, institution transport —9.8 per cent, and others —0.5 per cent. The average distance travelled was 2.5 km.

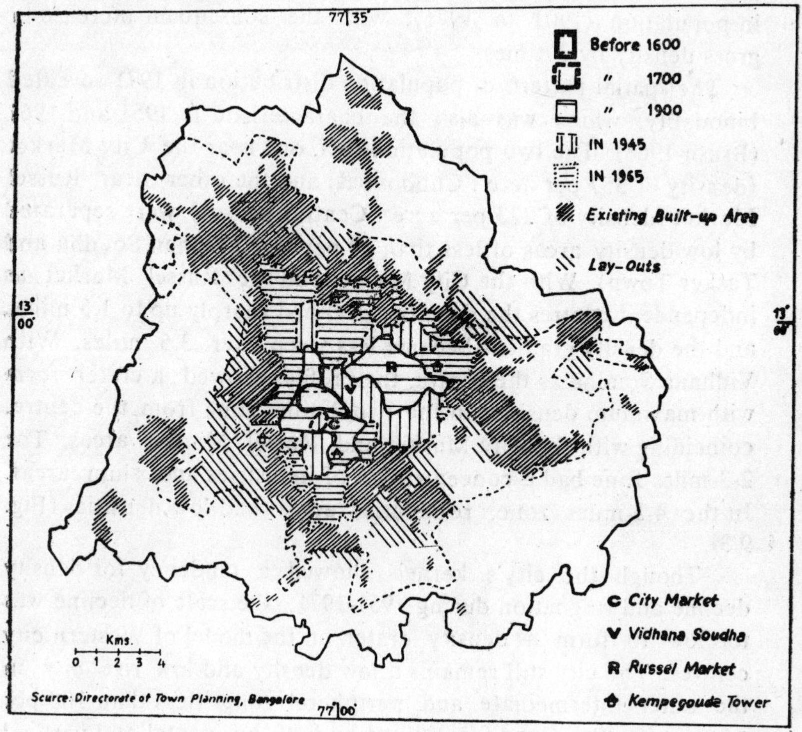

FIG. 9.2 : *Bangalore city—Urban spread*

TABLE 9.3 : Land use in Bangalore City, 1974

Predominant use	Percentage of the total area
Residential (excluding residential areas in defence and railway property)	37.5
Defence land (including residential development)	23.2
Agricultural and allied uses	10.6
Industrial	10.5
Park, play-grounds and recreational areas	8.0
Public and semi-public (including railway property)	7.6
Commercial	2.6

Source : Hanspal (1977).

area increase and population increase : 3 times in area and 9 ti mes in population (1901 to 1971), with the consequent increase in gross density by 3 times.

The spatial pattern of population distribution in 1971 revealed binodality, which was also the characteristic in 1951 and 1961 (Brush 1968). The two population foci, one near the City Market (density of 357 per acre : Cubbonpet) and the other near Russel Market (density of 323 per acre : Central Bazaar) were separated by low density areas of less than 50 per acre (Vidhan Soudha and Tasker Town). With the City Market and the Russel Market as independent centres the densities declined sharply up to 1.5 miles, and the density gradient became less steep after 3.5 miles. With Vidhana Soudha as the centre, the profiles showed a crater form with maximum densities in the 1 to 2 miles zone from the centre, coinciding with the City Market and Russel Market areas. The 2-3-miles zone had a concentration of industries and slum areas. In the 4-5-miles zone, residential areas had low densities (Fig. 9.3).

Though the city's kernels showed a tendency for density decline and stagnation during 1951-1971, the scale of decline was too low to form a density crater on the model of Western city centres. The city still remains a low density and low rise city in the outer intermediate and peripheral zones (less than 100 per acre) indicating scope for infilling by both horizontal and vertical growth.

THE CORE-INTERMEDIATE—PERIPHERY ZONATION

In this section, the spatial patterns in the city structure are examined with reference to the core, the intermediate areas and the periphery. The core consists of areas (census divisions in 1971) centered around the City Market, the Russel Market, and the Vidhana Soudha (an area that though was a part of the city in the early 20th century was developed as a central administrative complex only in the early fifties). The peripheral zone consists of all the census divisions on the periphery of the corporation boundary. The divisions in between the core and the periphery were designated as intermediate zone.

Sharp differences existed between core, intermediate and periphery of the city in terms of density, and residential housing

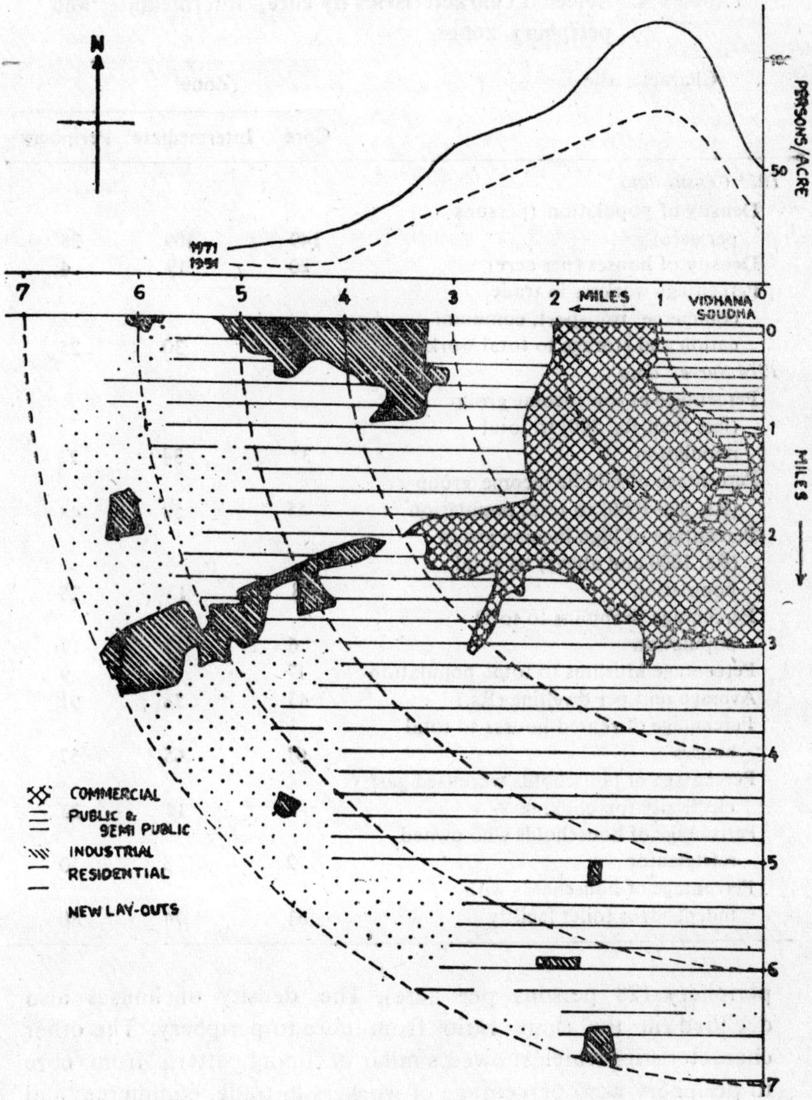

FIG. 9.3 : *Population density—Land use*

and living conditions (Table 9.4). The population density in the core (149 persons per acre) was more than five times that in the

TABLE 9.4 : **Selected characteristics by core, intermediate and periphery zones**

Characteristic	Zone		
	Core	Intermediate	Periphery
1971 Census data			
Density of population (persons per acre)	149	109	28
Density of houses (per acre)	20	19	4
Percentage workers in trade, commerce, transport, communication and storage to total workers	42	30	25
1974 Survey data			
Percentage of low income group (less than Rs. 300) to total population	37	34	27
Percentage of middle income group (Rs. 300-1000) to total population	55	53	48
Percentage of high income group (Rs. 1000 and above) to total population	8	13	25
Percentage Brahmins to total population	6	17	19
Percentage Muslims to total population	17	10	9
Average rent per dwelling (Rs.)	45	58	91
Percentage of rented houses to total houses	67	65	57
Percentage of households who used gas/electricity for cooking	7	14	25
Percentage of households who owned refrigerator	2	5	10
Percentage of households with independent toilet facility	61	74	80

periphery (28 persons per acre). The density of houses also declined in the same ratio from core to periphery. The other characteristics which showed similar declining pattern from core to periphery were percentage of workers in trade, commerce and so on, percentage of population in low income and middle income group, percentage of Muslims, and percentage of rented houses. The percentage of population in high income group, percentage of Brahmins, rent per dwelling, percentage of households who used gas/electricity for cooking, percentage of households who owned refrigerator, and percentage of households with independent toilet facility were higher in intermediate and periphery areas than in the core. While the city core acquired a special character by its centripetal pull and dominance over the periphery mainly due to the high concentration of work-places, financial institutions, higher order shopping areas, educational institutions, transport terminals, hospitals, and recreational and cultural centres, the periphery was characterised by higher levels of housing and living conditions than that in the core and intermediate areas.

FACTORIAL ECOLOGY OF BANGALORE CITY

Based on the household sample data (1974) and the 1971 census, 20 variables were selected to analyse the degree of spatial variations in the 63 sub-areas (census divisions) of the city. The selected variables were categorised into four groups : social, economic, family, and migrant status-tenure-density (Table 9.5). All the variables except density of persons and houses were taken from the household sample. The values were calculated as proportions of the total in the division. The distributions of the selected variables were then examined to find out the degree of spatial segregation of the households who share common characteristics.

Based on co-efficients of variation of the spatial distributions, the resident-migrant ratio (203%), proportions of the scheduled caste (150%) and Muslims (145%) showed a high degree of segregation in space, while average household size (15%) and sex ratio (18%) showed least segregation. The co-efficients of variations (CV) for proportions of Christians and Brahmins were 143 per cent and 113 per cent respectively, which also showed a high degree of segregation.

In the economic characteristics, the proportion of workers in high status occupations such as professionals, administrators, executives and related workers showed the highest segregation (CV : 82%). The least co-efficient of variation in the economic characteristics, 36 per cent, was in the distribution of service, production, and related workers. The income per household and per capita income, which indicate an overall mean income level, showed low degree of variation (37 per cent and 45 per cent respectively).

In the family characteristics, the spatial distribution of per cent woman workers to total workers showed the maximum variation (78%).

As a group, social characteristics showed maximum segregation, and family characteristics the minimum.

Both the Scheduled Caste and slum population showed the same generalised pattern of segregation and association in space especially near the mill areas,* though all scheduled castes do not stay in slums and all slums do not have only scheduled castes. Only the core areas which are essentially commercial areas and are highly congested have neither slums nor scheduled caste populations, just for want of residential space within the reach of this low income group.** Thus, the scheduled castes and slum population were concentrated in the peripheral and intermediate zones, and revealed interrupted zonal patterns.

At the other end of the caste-hierarchy scale, Brahmins were concentrated mainly in the older northern peripheries, all located in the new intermediary zones. Muslims were concentrated in the census divisions located in the core near Russel Market, while Christians were mostly in the north-eastern census divisions located in the intermediate areas.

The spatial distribution of administrative, executive, and managerial workers showed their concentration in areas in the northern and southern periphery. The other group which showed high coefficient of variation in the distribution was workers engaged in wholesale and retail trade industry and such workers were in high proportions in census divisions located in the core.

The areas of low socio-economic status with higher level of congestion and lower levels of urban amenities had more residents than migrants, and the high socio-economic status areas had low

* Industrial land use showed the maximum degree of segregation in space (coefficient of variation 204%). In sharp contrast to the high localisation of industrial land uses along the highways, industrial workers were highly scattered (coefficient of variation 27%) all over the city, resulting in easy movement from home to work place and back.

** The coefficient of variation of commercial land uses (93%) and workers engaged in trade, transport, storage, and communications (32%) indicated a spatial pattern of dispersal concentration, with maximum concentration in core, near both the foci.

TABLE 9.5 : **Co-efficient of variation in the spatial distribution of the selected variables**

Variables	Co-efficient of variation (%)
Per cent Muslim population to total population	145
Per cent Brahmin population to Hindu population	113
Per cent Christian population to total population	143
Per cent Scheduled Castes and Tribes to Hindu population	150
Per cent professional, administrative and related workers to total workers	82
Per cent clerical, sales and related workers to total workers	40
Per cent service, production and related workers to total workers	36
Per cent workers in manufacturing to total workers	42
Per cent workers in wholesale and retail industry to total workers	69
Per capita income	45
Income per household	37
Rent per dwelling	63
Household size	15
Females per thousand males	18
Per cent women workers to total workers	78
Number of children in age group 0-9 per thousand females in age group 15—49 (fertility ratio)	37
Persons per acre	83
Houses per acre	87
Per cent rented houses to total houses	29
Residents per hundred migrants	203

population densities, higher level of urban amenities, and a low resident-migrant ratio.

Of the 20 variables, per capita income was significantly correlated (positively or negatively) with the maximum number (10) of variables (Table 9.6), viz., per cent Brahmins to total Hindu population (0.53), per cent professional, executive and related workers (0.74), per cent service, production and related workers (—0.63), income per household (0.87), per cent dwellings (0.83), household size (−0.55), fertility ratio (—0.47), persons per acre (—0.36), and houses per acre (—0.37). This showed that in areas where per capita income was high, proportions of Brahmins, professional, executive and related workers, income per household, and rent per dwelling were also high, but proportions of

TABLE 9.6 : **Significantly correlated variables with selected variables**

Variable	Number of significantly correlated variables	Correlated significantly at 1% level with variables
1	5	4, 6, 8(—)*, 9, 20
2	7	4(—), 5, 6, 7(—), 10, 11, 12
3	1	4
4	4	1, 2(—), 3, 20
5	6	2, 7(—), 10, 11, 12, 16(—)
6	5	1, 2, 7(—), 9, 16(—)
7	7	2(—), 5(—), 6(—), 8, 10(—), 12(—), 16
8	3	1(—), 7, 9(—)
9	3	1, 6, 13
10	10	2, 5, 7(—), 8(—), 11, 12, 13(—), 16(—), 17(—), 18(—)
11	5	2, 5, 7(–), 10, 12, 16(—)
12	8	2, 5, 7(—), 10, 11, 13(—), 17(—), 18(—)
13	4	9, 10(—), 12(—)
14	0	—
15	0	—
16	7	5(—), 6(—), 7, 9, 10(—), 11, 17
17	3	10(—), 12(—), 18
18	3	10(—), 12(—), 17
19	0	—
20	2	1, 4

* Minus sign in parentheses indicates negative correlation.

service, production and related workers, household size, fertility ratio, and population and house densities were low. The variable rent per month which had significant correlations with 8 variables in the group of variables related with per capita income showed similar pattern of inter-relationships.

Among the variables, other than those of status, the variables which had significant correlations with at least five variables were : per cent clerical, sales and related workers which had significant correlations with Muslims (0.34), Brahmins (0.36), per cent workers in wholesale and retail trade (0.55), per cent service and related workers (—0.77), and fertility ratio (—0.33) ; and proportions of Muslims which had significant correlations with per cent scheduled castes (0.34), per cent workers in wholesale and retail trade (0.34), resident-migrant ratio (0.58), per cent workers in manufacturing (—0.43), in addition to its correlation with per cent clerical, sales and related workers. Another pair of inter-related variables which emerged from the analysis was that of Christians and scheduled castes (0.48).

Four underlying dimensions of variations (components) in the spatial distribution of 20 variables could be identified using Principal Component Analysis, which together explained 63 per cent of the total variation. The most important dimension (component one) explained 28 per cent of total variation of the total variation in the census divisions of the city, due to the 20 variables (Table 9.7). The other three dimensions of variation explained respectively 14 per cent (component two), 13 per cent (compo-ponent three), and 8 per cent (component four) of the total variation. Based on the mix of variables, the four dimensions were named as :

1. Socio-economic status
2. Religion-based segregation
3. Congestion
4. Household character.

1. *Socio-Economic Status* : This component represented the group of variables reflecting the socio-economic status. The variable of per capita income, which also showed significant correlation with a maximum number of variables, had the highest component loading (0.916). The other four variables with high positive loadings* were income per household (0.863), per cent professionals, administrative and related workers (0.823), rent per dwelling (0.803), and per cent Brahmins (0.706). The variables negatively correlated with the dimension of socio-economic status were per cent workers in service, production, and related occupations (—0.787), fertility ratio (—0.556), houses per acre (—0.498), persons per acre (—0.472), and per cent workers in manufacturing industry (—0.408). The component loadings or the correlations of the variables in this dimension showed positive inter-relationships among the variables of income, rent, and high

* The variables having component loadings of less than 0.4 are not presented in the tables though they were used in computing the component scores.

TABLE 9.7 : **Component loading of significant variables in the four components**

Variables	Component loading
Component 1 : Variation explained 28%	
Per capita income	0.916
Income per household	0.863
Per cent professional, etc., workers	0.823
Rent per dwelling	0.803
Per cent Brahmins	0.706
Per cent workers in services, etc.	—0.787
Fertility ratio	—0.556
Houses per acre	—0.498
Persons per acre	—0.472
Per cent workers in manufacturing	—0.408
Component 2 : Variation explained 14%	
Per cent workers in wholesale, retail, etc.	0.762
Per cent Muslims	0.749
Per cent clerical and sales workers	0.708
Resident per hundred migrants	0.562
Per cent workers in manufacturing	—0.496
Component 3 : Variation explained 13%	
Persons per acre	0.673
Houses per acre	0.628
Per cent rented houses	0.473
Per cent scheduled castes and tribes	—0.599
Per cent Christians	—0.569
Component 4 : Variation explained 2%	
Females per thousand males	0.621
Persons per households	0.528
Resident per hundred migrants	0.428
Per cent woman workers	0.398

status occupations and negative inter-relationships among the variables of low status occupations, fertility ratio, and densities. This further confirmed the nature of the component as one reflecting the socio-economic status of people.

2. *Religion-based segregation* : This component was next in importance to the component of socio-economic status because it explained 14 per cent of the total variation. The component mainly represented the high proportion of Muslims and the inter-relationships arising out of it, namely, the higher representation of Muslims and sales, clerical, and related workers. The variables

which had high positive correlations with this component are per cent workers in wholesale, retail, and related occupations (0.762), per cent Muslims (0.749), per cent clerical and sales workers (0.708), resident-migrant ratio (0.562).

3. *Congestion* : The third component which explained 13 per cent of the total variation represented mainly the density variables, namely, persons per acre (0.673) and houses per acre (0.628). The variable per cent rented houses had a positive loading of 0.473, confirming the finding of more rented houses in the core than in the intermediary and periphery zones in the section on residential housing and living conditions. The negative loading of the variable per cent scheduled castes and tribes (--0.599) confirmed the finding of inverse relationship of slums with density. The negative loading of per cent Christians (—0.569) showed that Christians lived largely in less congested areas.

4. *Household character* : The fourth component which explained 8 per cent of the total variation showed the inter-relationships of higher sex ratio, large household size, more residents, and higher proportion of women workers. The component loadings of these four variables were respectively 0.621, 0.528, 0.428, and 0.398.

The least component loading (0.398), which was the highest among the correlations of the per cent woman workers variable with the four components, showed the insignificance of the variable in the urban structure. It should be noted that in the studies of urban ecology for western cities per cent woman workers was an important variable in the dimension of "urbanisation" as identified in these studies.

Spatial patterns of the component scores

The component scores on each of the four components were grouped into "high", "high medium", "low medium", and "low" groups and were mapped to examine the spatial patterns of the dimensions of the ecological structure. The grouping was done using Ward's grouping algorithm based on the criterion of minimum increment to the sum of the squares (Ward 1963).

Component 1—Socio-Economic Status : The map of factor scores on the socio-economic status component showed high scores, ranging from 2.07 to 2.96, for three census divisions, viz.,

Kodandaramapuram, Palace, Sankarapuram* (Fig. 9.4). Thus, these three divisions emerged as the areas with households of the highest socio-economic status. The divisions which were grouped as high medium were Malleswaram, Seshadripuram, High Grounds, Jayanagar, Wilson Garden, Yediyur, Basaveswara Temple, Visveswarapuram, and Murphy town. Thus, except for Murphy town, all the divisions with high and high medium scores

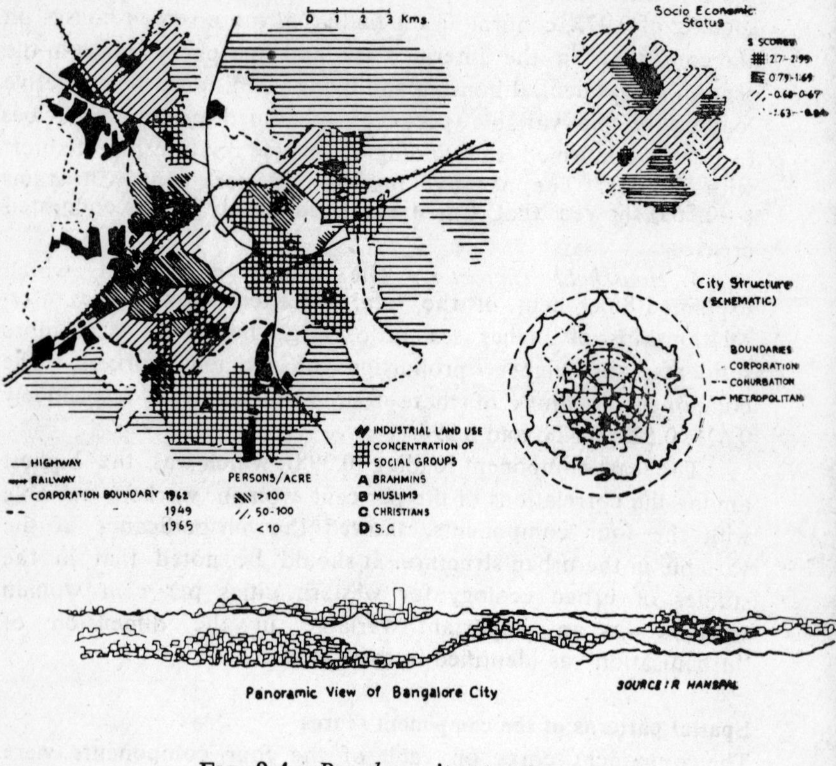

FIG. 9.4 : *Banglore city structure*

on the socio-economic component were in the central, north, and south sections of the intermediary and periphery zones. An examination of the distribution of households by income classes

* In the maps of factor scores the division 16 (Chickpet) was left blank as no sample block for household survey from the division got selected in the sample and the division was not included in the analysis.

in the divisions revealed that while the divisions with high scores had 43 per cent to 67 per cent households with per capita income of Rs. 200 and above, the divisions in the high medium group had 26 per cent to 57 per cent of such households.

The divisions in the group of low socio-economic status were concentrated mainly round the two nuclei of the core—the City Market and the Russel Market. These divisions were Tasker town and Broadway in the east near the Russel Market and Dharmaraya Temple, Anjaneyaswamy Temple, Cubbonpet, Subashnagar, Kempapura round the City Market, and in the western part of the city. The other divisions in the low group were Maharaja Mills, Super Talkies, Sriramapuram, Bhashyamnagar, Ramachandrapuram, Swimming Pool Extension.

All the other divisions except those mentioned above were in the low medium group and were mainly in the east and west part of the periphery.

Component 2—Religion-based segregation : Muslim segregation arises from an apparent need for proximity to co-religionists. The socio-economic status which forces Hindu households of differing status to live in separate areas has no relationship with the distribution of Muslim households. This was also reflected in the factor loadings of variables per cent Muslims (0.749) and per capita income (—0.012) and correlation coefficient between them (—0.017), showing that the spatial distribution of per cent Muslims had no relationship with the distribution of per capita income, the most important variable of socio-economic status. Gist (1957) also found that "of the religious minorities in Bangalore none is perhaps as highly segregated as the Muslim community".

Component 2, which reflected high proportion of Muslims, sales and clerical workers and residents showed high factor scores for the divisions Broadway and Maclver town and high medium for the divisions Central Bazaar, Tasker town, Frazer town, Kalasipalyam, City Market, Bhashyamnagar, and Malleswaram. The low group in this component showed a high proportion of workers in manufacturing.

Component 3—Congestion : Three clusters of divisions with high and high-medium scores emerged from the map of factor scores on this component ; (i) Bharatinagar, Commercial Street, Ulsoor, Swami Setty Garden, and Shivajinagar in the east ; and (ii) Seshadripuram, Bhashyamnagar, Malleswaram, Sriramapuram, Dattatreya Temple, (iii) City Market, Rangaswamy Temple, Cubbonpet, Anjaneyaswamy Temple, Kasivisveswara Temple, Dharmaraya Temple, Chamarajpet-West and-East, Fort, Kalasipalyam, Visveswarapuram, Sankarapuram, Basaveswara Temple, Mavalli, and Kempapura Agrahara in the west. These divisions had high density of persons and houses and high proportion of rented houses.

The divisions with low scores on this component were Tasker town, Austin town, and Murphy town in the east and Maharaja Mills and Padarayanapura in the west. These showed less congestion and larger proportions of Christians or scheduled castes or both. The rest of the divisions were grouped in the low medium group.

Component 4—Household character : This component, positively associated with females per thousand males, household size, resident-migrant ratio, and per cent woman workers, had high positive scores for three divisions : Broadway, Shivajinagar, and Maharaja Mills.

PATTERNS OF THE CITY STRUCTURE

The foregoing analysis showed that the city's structure was found to be not a simple "rich centre and poor periphery" but a complex one : middle class status city centre and both high status and low status periphery.

The outward moving periphery left behind older peripheries and constituted a distinct intermediate zone, not only by its intermediate geographical position between core and periphery but also by its intermediate and mixed position in the ecological scale. Regarding the tripartite, the polarised socio-economic groups were multi-nuclear in distribution ; concentric zonation was not mappable in simple geometrical forms, but the dominance of the city core and distance from the city core were found to be significant in socio-economic zonal patterns. Structural patterns did not emerge clearly, not because of their absence but because of the insensitivity of the data. The district corridor pattern of mixed structure proved one strong evidence of this influence. The overall structure of the city revealed both east-west zonation, more religion based, and core-intermediate-periphery zonation based on socio-economic variables.

TEN

JAY A. WEINSTEIN

SUBJECTIVE COMPONENTS OF URBAN ECOLOGICAL STRUCTURE

A Theoretical Statement and an Application to Two Indian Cities

THIS paper is about the components of urban ecological structure classified, broadly, as "subjective". The focus is on the interaction between some of these components and the more familiar objective features of urban ecological structure such as physical distance, location of transport routes, land uses, and land values. The approach outlined, and applied in a limited manner, has roots in the landmark work of Firey (1945).

Firey's revision (or, perhaps, "rejection" would be a better word) of the classic ecological models was one of three major innovations in urban sociology in the years immediately following World War II. Of the three, it is today the least well known and least well developed. While some of the reasons for this lack of attention to Firey's subjective (or "symbolic" or "cultural") ecology are suggested in the following discussion the principal purpose is not to recreate the debate between subjectivists and objectivists, nor to seek the sources of evolution of urban ecology (as an historian of ideas or a sociologist of knowledge might). Rather, the discussion is confined to three specific tasks : (i) an overview of the development of urban ecology's interest in subjectivity during the past thirty years ; (ii) a brief catalogue of insights to be found in the geographical literature specializing in the related study of "subjective space" or "mental mapping" ;

and (iii) application of some of these observations to the interpretation of data on two contemporary cities in India—Madras and Ahmedabad.

With a few major exceptions, sociological studies of urban ecological structure in the United States have made little use of Firey's insights or of the subjective approach in general. Instead, the concerns of the two other schools which were first proposed at about the same time as Firey's have dominated theoretical formulations and research designs in urban studies—and in other sociological specialities as well. But this is not to say that no social scientific research of the subjective component has taken place. In addition to the exceptions within sociology, to be noted, geographers and urban planners have explored this area in some depth and detail.

By introducing cases from India, a country which was not studied from an urban ecological perspective until the late 1960s, a suggestion is made here that it is perhaps necessary to examine cities in a society different from that of the United States in order to appreciate the value of the subjective approach. To summarize an argument that is presented in greater detail below, given the fact that the subjectivity of actors, sociological observers, and audiences in the United States is, fundamentally, shared, urban ecological studies have progressed fairly well with no need to resort to special, subjectivity-oriented terms or explanations. When attempting to apply the objective models developed in the United States to cities in a society as different from the United States as India, however, the need for—and the value of—subjective ecology is apparent.

The class of subjective components discussed here is large and diverse ; it includes a wide range of types of social-psychological states and dispositions attributable to (members of) a city's population. To suggest that none of these has been incorporated into urban ecological theory or research would be to overstate the case. Even the most objectively-oriented urban ecologists recognize the importance of subjectivity as a determinant (or at least a consequence) of a city's distribution of social aggregates and land uses. Lacking in most sociological writing on urban ecology, however, is a clear statement of : (1) which of these subjective components (or psychological states) have particular bearing on the study of urban ecological structures ; (2) what is the relationship—that is, in terms of cause and effect—between subjective components (as a class or individually) and the objective components ; (3) what bearing does knowledge of the subjective components have on research and theory: Does it improve the predictive power of our models ? Does it increase the usefulness of the ecological perspective to planners, engineers and administrators ? The following discussion of past research in urban ecology and geography and of work on Indian cities is intended to provide a partial statement of these three principles.

THE SUBJECTIVE APPROACH IN SOCIOLOGICAL LITERATURE

Revisions of the post World War II era

The classic models of the socio-geographic organization of cities introduced by Burgess (1925) and Hoyt (1939) have been subjected to empirical test or rejected hundreds of times in sociological, geographical planning and related literatures by a group of schools (for recent summaries, see Schwirian 1974, Berry and Kasarda 1977, and Gist and Fava 1974, Weinstein 1975). One of these schools, that first elaborated by Firey in the late 1940s, contains a powerful critique of the classic models and a sound set of directions for improving upon them. In order to characterize Firey's school one must focus on the era in which it originated, that is, 1947-1950. During this period a series of works was published that summarized the ecological research of the preceding two decades and which continue to inform (and divide) the field today.*

At the latter end of this period came Hawley's (1950) and, later, Duncan's, Schnore's, Gibbs', and Martin's contributions to what Willhelm (1964) terms the neoclassical school. With this work, the exclusive spatial focus of the classic approach was abandoned as being geographically determinist and atheoretical. Instead, emphasis was placed on the systematic character of aggregate human relations, independent of the configurations that these systems of aggregates might trace on a city map. Hawley

* According to Willhelm (1964), Suttles (1972) and Berry and Kasarda (1977), the post-World War II critique of classic ecology (including the work of Firey) can be traced to the research of Alihan (1938) and Gettys (1940).

(1968) has delineated the role of space (and time) in the neoclassical approach in this way :

> Although the emphasis is centered on the functional system that develops in a population, it is not intended to exclude concern with spatial and temporal aspects ; rather, these aspects are regarded as useful dimensions for the measurement of organization.

The express concern of Hawley and the other neoclassicists was to strengthen the connections between human ecology and plant and animal ecologies. As a result, the system metaphor employed by the Chicago School, Hoyt and their students was challenged as being naively biologistic (and Darwinian at that). In place of the invasion/succession/competition and related subprocesses stressed in the earlier system metaphor, Hawley and his colleagues favoured a physically-oriented, moving equilibrium set of subprocesses (a "cyber-system" in current language).

An earlier major landmark of the neoclassical approach is the P.O.E.T. mnemonic (Duncan 1959).* Among its more recent achievements is a virtual domination of American sociological studies of social stratification and mobility (now called "the status attainment process". See Tyree and Hodge 1978). The evolution of the neoclassical approach has led human ecology increasingly further from the more easily identifiable features stressed in the classic models. Yet, it is from the classic models that this, now general, orientation is derived (See Horan 1978).

A second contribution of the 1947-1950 period is the development of social area analysis by Shevky and associates (Shevky and Williams 1949, Shevky and Bell 1955). This approach viewed the classic models as useful but as basically

* I.e., P=population, O=social organization, E=environment, T=technology. Gist and Fava (1974) amend this to include "SP"=social psychological factors. This amendment does indeed bring the neoclassical approach closer to that of Firey, but it is not necessarily in keeping with the orientation of Duncan and Hawley. I am inclined to agree with Willhelm who argues that the neoclassicists "generally define the responses of man's sustenance and/or social organization to certain 'objectified' elements as the domain for ecological investigation ; they firmly reject notions of subjectivity" (1964).

unsubstantiated by empirical test. Some of the concepts of Burgess, Hoyt and their students were operationalized by Shevky, and certain of the propositions were put to a test (first in Los Angeles and San Francisco, later in scores of cities in the U.S. and abroad.*

Social area analysis provides a classification scheme describing the sociogeographic structure of an urban population. Spatial aggregates, usually census tracts, are assigned scores on each of three social area indices (traditionally calculated from census variables) : social rank, familism and ethnicity.** Tracts (or, theoretically, neighbourhoods) of similar rank on a particular one of these indices are considered to be occupants of the same "area" of social space. The three social area configurations are then referred back to a map in order to determine the extent to which social area corresponds, in a systematic fashion, with *natural* area.

Since the pioneering work of Shevky, this school and its variants (including factorial ecology) have occupied a prominent role at "the cutting edge" of urban ecological research. Bell (1955) first subjected the question of whether or not Shevky's indices (and their ordering) were in fact naturally-occurring dimensions. Employing a factor analysis of the same variables that Shevky had used to construct his indices, Bell verified that the three-part classification scheme did emerge from the census data as analyzed. Anderson and Egeland (1961) and Anderson and Bean (1961) first showed that the geographic distribution of the social areas (now operationalized as factor scores) corresponded to the classic spatial models, that is, social rank distributes in a sectoral manner, family structure characteristics in a concentric zone fashion, and ethnic minorities (following Harris and Ullman

* Schwirian's (1971) excellent methodological paper summarizes the principal weaknesses of the approach. Schwirian is, however, a "critic from within" the social area school.

** Note that each index—or dimension—represents a particular mode of status assignment : social rank—achievement, familism—acquisition, ethnic minority status—ascription. Shevky hyothesized that this separation—and the dominance of social rank—was characteristic of modern, urban social structure. This hypothesis is the basis of the series of studies in which urban ecological structure is used as an indicator of modernity. See Abu-Lughod (1969), Berry and Rees (1969), Weinstein (1976b) and Weinstein and McNulty (1979).

1945) in multiple nuclei scattered at various points in the city. While several revisions of these conclusions have been suggested (e.g., Van Arsdol, Cammilleri and Schmid 1958, Schwirian and Matre 1974), the basic social structural and geographic configurations first discussed by Shevky have been shown to characterize the major cities of the United States and, as variants of a general model, of Canada and Europe (see, e.g., Sweetser 1965a) as well.

Thus, while Hawley and the neoclassicists de-emphasized the spatial component and rejected the biological metaphor inherent in the classic models, Shevky and the social area analysts initiated the practice of subjecting the models to empirical test (and, as a result, of verifying them in application to individual cities throughout the U.S. and elsewhere). The third innovation in ecology of the post World War II period (the first of the three in time) is the subjective approach of Firey (1945) and Hollingshead (1939, 1947).* This strategy combines elements of both the neoclassical and social area critiques of the Burgess and Hoyt models, yet it introduces a set of elements—the subjective component—which neither the classic ecologists nor their critics had considered.

In Firey's work on Boston (Firey 1947) the theoretical discussion and the research design are focused on "explaining the territorial arrangements that social activity assumes". This programme, so stated, is virtually the same as that of the classic ecologists (and of the social area analysts). Yet it differs in emphasis from the neoclassical approach which "seeks knowledge about the structure of a social system and the manner in which that structure develops" (Hawley 1968). At the same time, Firey and Hollingshead agreed with the later critics, that by World War II urban ecology had "collapsed into a theoretical map-making and description (Berry and Kasarda 1977)". In light of this assessment, Firey chose to provide the map makers with a

* Hollingshead was criticized by Firey (1947) for his economic rationalism. This critique was based, however, on the work of Hollingshead that was accessible to Firey at the time, that is, Hollingshead (1939). While *Land Use in Central Boston* was in press, however, Hollingshead's (1947) re-examination of ecological theory appeared. This article offers a revision of the classic school which is entirely within Firey's concept of subjective ecology.

theory tracing the connection between human consciousness and the geographic distribution of social activities.

The outlines of subjective ecology

Briefly stated, Firey's (1945, 1947) argument attempts to explain the distortions of the classic models of land use, land value and geographic distribution of social groups observed in Boston and elsewhere. That such distortions exist, in general, appears to be beyond doubt to Firey, although his only exact evidence is drawn from Boston.

His review of previous research reveals that three types of theories have dominated ecological studies: (1) idealized descriptive schemes (which include the concentric zone and sectoral models); (2) empirically rationalistic theories; and (3) methodologically rationalistic theories. Each of these types seeks to explain the sociogeographic character of a city or section of a city without reference to cultural factors such as values, sentiments and group traditions. Instead, the theories employ one or another version of a causal explanation which treats human consciousness as entirely passive.

Those of the first type view the formation of urban ecological structure as a self-regulating mechanism.

> Nowhere in the theory is there a definite statement of the *modus operandi* by which people and groups are propelled to their appointed niches in space. Yet the whole emphasis upon the inevitability of the pattern implies a disvowal that they have any active part in the locational process . . . at the very best (Burgess) concedes that cultural factors may exert a 'resistance' . . . (Firey 1947).

Theories of the second type, the empirically rationalistic, share an orthodox economic view of consciousness. That is, to one degree or another, they treat human actors as ultimately rational calculators of locational costs and benefits who respond to time/distance constraints which are "given" in the structure of the market. The earlier work of the classic ecologists Hollingshead (1939) and McKenzie (1933) exemplify this approach for Firey in its insistence that ecological structure be explained as an economic system in equilibrium. Typical of this approach is

McKenzie's postulate—which Firey quotes (1947) : "Underlying all forms of urban segregation are the factors of income and rent" (McKenzie 1933).* In sum, "The rationalistic ecologists have treated physical space entirely as a productive agent which can variously be combined with other productive agents" (Firey 1947).

The third type of theory, the methodologically rationalist, categorically excludes non-rational factors in explaining urban land use in the interests of "pure" theory. These theories "far from claiming any descriptive accuracy, deliberately "fictionalize" the process of spatial adaptation" (Firey 1947). The outstanding example of this type of theory, according to Firey, is the classic work on location of Weber (1929). In Weber's theory, factors such as labour costs, accessibility of transport routes, and location and composition of raw material inputs are employed to explain the location of industries in a capitalist economy. Weber allows that he calls "cultural factors" (actually these are the historical conditions which contribute to regional and national variations in the wealth of capitalist enterprises) only to determine labour supply and costs. But neither Weber nor his critics (e.g., Oskar Englander and Andreas Prehohl) are willing to permit any subjective elements to enter their theories.

Each of the three types of ecological theories discussed by Firey is viewed as unrealistic in comparison to an approach, traced to Durkheim (1915), Sorokin (1943), and Simmel (1923), in which space is treated as a cultural entity. Absent in views which assume an intrinsic society-space relationship is a role for "non-intrinsic" values and interests which are known to affect the spatial activities of human actors and groups. Values (or volitional adaptation), for Firey, "comprise one of the criteria by which social systems (from individual to society) choose locations .... Interests (rational adaptation) dominate the spatial adaptation of certain social systems but these interests themselves

* Firey divides the category of Empirically Rationalist theory into "literal" and "empirically compromised" rationalists. The second group includes McKenzie and the Chicago School ecologists ; the first is illustrated with discussion of an article by Quinn (1940) on locational analysis.

come indirectly from broader and large cultural systems ; hence they cannot be viewed as self-given ends" (Firey 1947). The programme of Firey (1947), and of subjective ecology in general, is to integrate values, interests and the larger cultural systems they entail with objective components in explaining urban socio-geographic organization.

> What all this means for ecological theory is clear. If space gets its socially relevant qualities through cultural definition, and if social systems (when contingent on space at all) may locate in terms of social values there is obviously far more to the society-space relationship than an intrinsic nexus. A satisfactory ecological theory must somehow conceptualize this 'non-intrinsic" nexus and integrate it with the propositions applying to intrinsic spatial adaptation (Firey 1947).

In connecting the cultural emphasis in European sociological theory with the then (and still) current concerns of spatially-oriented ecologists, Firey provided future generations of urban researchers with a rich source of insights. The major portion of Firey (1947) is devoted to exploring some of these insights in a social-historical examination of four neighbourhoods : the South End, Back Bay, Beacon Hill and the North End. In this examination, a wide range of data types are employed to discern and illustrate the logico-meaningful components of land use and population distributions in these neighbourhoods. Information from newspaper reports, social registers, city directories, census and municipal surveys, records of land transactions, planners reports, and so on, is used to demonstrate the ways in which variegated histories, collective values and group interests have shaped the contemporary map of Boston. From this research and theoretical discussion, several specific relationships between culture, subjectivity and space are made explicit or implied in citations.

Later developments in urban ecological research (especially those related to the rise of the neoclassical and social area approaches) tended to exclude most of the insights of Firey and Hollingshead. Instead, the major impact of subjective ecology was experienced in the study of community process. The subjectivists had stressed, in opposition to the three types of theory which Firey criticizes, the effectiveness of community values in determining land uses and land values. This was illustrated in an especially dramatic way in Firey's studies of the ethnic communities of Irish, Italian and Black Bostonians who settled in the central city.

> Firey points out that sentimental and symbolic attachment to spatial location is not restricted to historic or upper class areas. The foreign-born Italian immigrants tend to remain in the North End of Boston even when their means permit them to move elsewhere, for the North End provides the traditional Italian community life not available in other sections of the city (Gist and Fava 1974).

This insight can be traced to the principal theoretical interests (and the site) of Whyte's study of black neighbourhood community (Whyte 1943). In conjunction with Whyte's study, the subjective approach has influenced Gans' (1962) research, that of Suttles (1968) and, directly or indirectly, the entire rich literature on neighbourhood community studies which was accumulated during the late 1950s and 1960s.

In most of these studies, however, the impact of values and interests on spatial distribution is not of central concern. Instead, geographic factors are generally treated as contextual. The focus of these studies, like that of the neoclassical ecologists, is on community process *per se* (in its aspatial and spatial manifestations). The advantages of the community focus over the neoclassical approach are evident. While the latter has, in principle, avoided treating human consciousness as an active force in community formation and maintenance, the former has explicitly focused on this factor. Nevertheless, both approaches leave unaddressed certain issues related to neighbourhood formation at a macro scale : community studies because their focus is generally microcosmic, neoclassical ecology because neighbourhood formation is, at best, of incidental interest.

While most ecologists, after 1950, pursued either a neoclassical or a social area approach, a few studies which directly extended the observations of Firey and Hollingshead did appear. This body of literature, though small and infrequently cited in ecological research, contains numerous key observations on the subjective component in the distribution of spatial activities.

Among the earliest of these studies (also cited by Willhelm 1964) are articles by Form (1954) on the determination of land use patterns and Jonassen (1954) on the values which underlie ethnic segregation. Sjoberg (1960) explored the impact of feudal versus modern values on urban land use patterns. Of special interest in Sjoberg's study is his focus on the evolution of specialized commercial and residential areas as a function of modernization (also see Abu-Lughod 1968d). While Sjoberg's classification scheme has been criticized for being overly simple (see, e.g., Cox 1969), his inclusion of cultural elements in a comparative perspective has special relevance for current cross-cultural ecological studies, as discussed in a later section in this paper.

The first complete elaboration of the subjective approach since Firey (1947) is to be found in Willhelm and Sjoberg (1960). In this work, attention is focused on the value and interest differences which arise over urban zoning decisions. In Willhelm (1962) these observations are expanded in a full scale study of zoning in Austin, Texas during the late 1950s. Following Firey's approach to data collection, Willhelm relied extensively on public records and interviews with relevant actors (planners, administrators, land owners, etc). He demonstrated that the actual land use patterns which developed in Austin were the result of political compromise, accommodation and a (perhaps unstable) resolution of conflicts between economic versus protective, collective versus individual and present versus future oriented values. Firey's concept of "proportionalization of ends" was found by Willhelm to provide a far more potent predictor of land uses in Austin (as it did in Boston) than any of the deterministic concepts of the classic ecologists.

Willhelm's observations certainly provided a challenge to the major schools in ecological sociology which were beginning to dominate the field at the time of the appearance of his book. Yet, the ecological literature of the later 1960s and early 1970s shows little acknowledgement of the contributions of Form, Sjoberg, Willhelm (or even Firey). Instead, the research techniques of the social area approach were increasingly refined and the number of studies demonstrating the fundamental utility of objective factors multiplied. Only quite recently has the focus on the role of values in determining socio-geographic distributions again come to the attention of sociologists. A paper on the distribution of black neighbourhoods in Chicago (Berry and Kasarda 1977), for example, is quite close in concept and conclusions to the studies of Jonassen and Willhelm. But, aside from these recent exceptions, the subjective approach in urban ecology has been conspicuous by its absence.

One last sociological study which deserves mention in connection with the subjective approach is Suttles (1972). Suttles' focus is primarily on community formation processes, as detailed in his earlier work (Suttles 1968). Yet, at various points, he addresses an aspect of the culture/space relationship which Firey had traced to Sorokin and Durkheim but which was not explored by later subjective ecologists. This is the cognitive dimension of urban geography or "mental maps".

A thorough application of the European cultural perspective on space would, according to Firey, include not only group values and interests that bear on land use decisions, but it would also relate to "larger and broader cultural systems" including cognition of space relations. If Sorokin and Durkheim were to be taken seriously about the cultural determination of space, then subjective ecologists would explore the ways in which subjectivity influences the social perception of distance and location (factors which have been treated as wholly objective in the more orthodox research in contemporary urban ecology). Suttles, among all of the community-oriented urban sociologists, has attempted to deal with cognition of space in a theoretically integrated manner. His principal point in this regard is related to his general observations on the functions of community affiliation, especially among the urban poor. One basic community type, the defended neighbourhood, has particular bearing on how residents view location and distance. Like other forms of knowledge, cognition of space serves to distinguish and protect members of the in-group from those of the out-group. The function of cognitive maps in the defended neighbourhood is to :

> provide an internally consistent set of directives, a collection of consensually shared understandings about what one's

> spatial location means, and a separation beween persons apt to fall into conflict (Suttles 1972).

Cognitive maps are associated with other community types beyond the defended neighbourhood including the "community of limited liability" and the "contrived community". For each of these types, related sets of social priorities and stipulated physical boundaries combine to influence the manner in which community residents perceive their environments. Especially relevant in light of the work of Sjoberg and Willhelm are Suttles' views on the impact of *planning* on cognitive maps.

> In the past, the physical structure of most American cities and the cognitive maps which people imposed upon these cities have tended to follow rather independent courses . . . . Now, with the advent of urban planning the physical structure of the city is no longer subject only to economic process but involves politics and some cultural image of what the city ought to be like (Suttles 1972).

Thus, in Suttle's work, the cognitive and value/interest components of ecological structure are integrated. As his argument develops, however, these components are employed in a broader discussion of the process which gives rise to limited communities (in the face of the perennial quest of the "excessive vision of local community") in urban America. No attempt is made, for example, to classify types of mental maps, to operationalize the concepts involved or to apply them in survey research. Notwithstanding these limitations, Suttles is the first sociologist since Firey to explore in detail the cognitive aspect of subjectivity in ecological process and structure.

With this mention of Suttles, the overview of the subjective approach in sociological studies of urban ecological structure can be summarized. Firey and Hollingshead, following earlier critiques of the classic models, indicated that the cultural determination of space had been overlooked in previous research. Firey, in particular, discussed and illustrated the many ways in which values, interests and related symbolic aspects of urban life bear on the determination of land use. Following upon these observations, Willhelm and others focused upon the value conflicts which informed decision-making processes in zoning and planning. Other aspects of Firey's revision were explored in the context of community studies rather than within ecological research *per se*. Some of these studies, especially those of Suttles, did return to the cognitive map orientation implicit in Firey's theoretical foundations. Lacking to this day, however, in most sociological work on ecological structure are (1) an extended treatment of subjective determination in general, and (2) the development of theory and methods for using the cognitive map perspective in empirical research. In order to get a more complete view of the latter direction, it is necessary to turn to the work of geographers.

A CATALOGUE OF GEOGRAPHIC RESEARCH ON MENTAL MAPS

The mental map or subjective space approach in geography is a branch of behavioural geography. Its origin has been traced by Gould and White (1974) to work done by Trowbridge (1913). Trowbridge was one of the first geographers to investigate systematically the problem of "sense of direction". He noted that the ability of people to navigate in a strange city varies between individuals. From this observation, he postulated a process wherein people more or less successfully orient themselves "egocentrically". Thus, when highly egocentric individuals move about in a strange place, they maintain their orientations by rotating their mental maps about the centre. Buttimer, in an outstanding review of continental literature, indicates that the subjective space concept can be traced to a work by Sion (1909). In this study of peasants and their relationship to the land in Normandy, a question is posed which could as well have been put by the later subjective ecologists : "should we be satisfied with drafting an opaque, objective map of social patterns in space, or must we supplement this with the subjective or inside view' ' (Buttimer 1968).

From these beginnings, the role of subjectivity in spatial behaviour has been explored in a great variety of ways by geographers. Downs (1970) identifies three main areas of concentration. The first takes the form of a structural or "cognitive mapping" approach *per se*. In this approach the geographer/psychologist attempts to identify and map out the individual's space perception. The rationale for this approach is the view that people's perception of the environment affects their spatial behaviour, that is, what they do with space and what they do in space. The second approach is the evaluative. This area is concerned with the way in which people perceive their environments through spatial images and subsequently evaluate these images through a decision-making process. These images are aspects of the environment that people perceive as important and use as a basis for structuring their behaviour within it. The third area is the preference approach. This field is concerned with the way in which people assess a set of spatially differentiated areas on a "scale of preferences". This approach is similar to the evaluative except that most of the research that uses this approach focuses on the analysis of intra-city migration. These three approaches attempt to understand the nature of perceived images, their relation to the "real" world, the processes which influence the transformation of the "real" world into images, and the effects of the image on the decision-making process underlying the interaction between people and environment.

Cognitive mapping

The modern cognitive mapping approach had its beginnings with Kevin Lynch's study (Lynch 1960). The mapping approach deals in general with how people perceive their objective surroundings. The methods used by geographers to evaluate these perceptions usually rely on maps drawn from memory or other techniques to transform perceptions into paper and pencil maps. It is assumed that people form conceptions of their significant environments and that these conceptions are formed in a systematic manner. Stea (1969) has suggested five sets of parameters by which people order their perceptions :

1. Establishment of hierarchies ; some areas are more important than others
2. Boundedness ; the space conceived ends somewhere
3. Objects and places are located within the space and can be conceived as points more or less clearly defined.
4. The points exist in some relation to each other, that is, one can speak of :

(a) Distance between points
(b) Bearings from one point to another
(c) Routes from one point to another which include the changes in direction taken in following the route

5. Connectedness between points :

(a) A route existing between two points implies connectedness
(b) Any interference with the route such as lack of transportation or any formidable interference constitutes a barrier.

The maps that are formed by individuals are not maps *per se*, but are functional analogues of what people perceive in their minds. The paper and pencil map which subjects reproduce is a cognitive (that is logico-meaningful) representation much like what a cartographer tries to represent in a physical map : an image of "the world in our heads". According to Downs and Stea (1973), "Cognitive mapping is a process of acquiring, amalgamating and storing spatial information". Thus cognitive maps are seen as ways of systematizing or organizing subjective space ; but, clearly, there is also a process of selection involved. Stea (1967), for example, speaks of "invisible land scapes" which are physically present but which do not play a part in perception. Some maps are incomplete, while other maps are distorted and contain "significant" areas for different people. For some people fantasies play a role in forming their mental maps ; other maps are constantly changing as new information is gathered. These invisible landscapes catalogued by Stea are as follows :

1. Attraction areas ; always present and rarely or never utilized or viewed
2. Attraction areas ; utilized or viewed, but seldom present
3. Deterrent areas ; always present and rarely or never viewed or utilized
4. Deterrent areas ; utilized or viewed, but seldom present.

While much of the geographical research on cognitive mapping has focused on underlying psychological processes (for example Gould 1967, 1968, 1969), several recent studies have related directly to ecological issues. Clark (1969), for example, found that areas nearest to one's residence are more prominent in cognitive maps. Repeated events such as journey to work, shopping and personal recreation are most important in building cognitive maps ; thus one's map is most distinct in familiar areas. Geographers believe that this affects one's use of "personal space" and is an extension of his personality and culture (Sonnenfeld 1968). But this map of visual landscapes and available services is not necessarily rationale ; as "satisfying behaviour" it is affected by biases, prejudices, and personal experience. In behavioural terms it is

> " . . . a series of psychological transformations acquired, coded, stored, recalled, and decoded information about relative locations and attributes of phenomena in man's everyday spatial environment" (Downs and Stea 1973).

Thus, these maps are a basis for decoding and implementing any strategy of spatial behaviour or everyday life experiences within the physical environment.

To test the assumptions of diverse perceived worlds, Orleans (1973) drew a sample of respondents based upon an application of the Shevky social area analysis formulation in Los Angeles, and his findings shed some light on how subjective components can be integrated with more orthodox variables. Orleans found that low social rank areas showed an "attenuated world view, frequently limited to an image of the city extending only a few blocks", while higher social rank areas showed a more worldly view. In Orleans words,

> " . . . discrepancies between maps by respondents' different social worlds which may touch, but in many instances do not interpenetrate."

To this may be added the findings of Golledge, Briggs, and Demko (1969) in which they found two sides to the perception of urban space, an objective and imagined space. They used this distinction in the analysis of distance. Their findings showed, for the first time in empirical research, a propensity to perceive distances as longer when urban space is denser (central business district). Similarly, Michelson (1971) broke distance into perceived distance and accessible distance. The perceived distance between two points becomes shorter when the distance between these two points is "... symbolically conducive to being transversed". This observation supports the findings of Suttles and Clark concerning familiarity with one's home location, that is, degree of perception (or preference) becomes stronger in areas that are more accessible to the individual, the person's home location. Thus distance between points (both perceived and real) may affect the interaction between people residing at these points.

Other observations on cognitive mapping in the geographic literature tend to focus on technical and epistemological issues. While these, too, are of interest to subjective ecology, they are of less direct relevance than those already cited. The general theoretical contribution of the cognitive approach lies in the many ways indicated in which space is treated as a social-psychological construct. Firey's subjective perspective drew attention to the fact that space—for a sociologist—is not a material, immutable even, that it, too, is shaped by the subjectivity of actors. Thus, to treat space only as an independent variable is to distort it and its effects on social activity. While Willhelm and others have elaborated upon the subjective, cultural factors which influence land use, and so on, cognitive map-oriented geographers have focused on the subjective factors which shape space itself. Most of these observations are, to be sure, couched in an individual psychology language (for example, group differences in cognitive mapping are generally not treated as central). They can, however, be seen as readily adaptable to a social-psychological ecology in light of Firey's contributions.

Evaluative and Preference Approaches

The evaluative approach is closely related to the preference approach in that both take into account the role of values in forming subjective images of the environment. They differ from the cognitive mapping approach in placing more emphasis on the "imagined" perception than on the "objective" world (although, from an epistemological stand point, the two may both be subjective). The evaluative approach concentrates on the normative representation of space. That is, the individual is viewed as appraising his/her perception of the environment. This perception is

> " . . . affected by the values which people put on the places in their different locational pigeon holes" (Cox and Zannares 1973).

The environment that is perceived may be classified into distinct "niches", that is, different areas denote different feeling states. Cox and Zannares formulated a classification system by which areas are assigned properties based on perceptions. They hypothesize that these classes organize perceptions. The first level, a hierarchical structure, classifies regions from largest to smallest (nation-state-country-city-neighbourhood). The next system classifies areas by certain "internal locational characteristics", i.e., it distinguishes between city-country-suburb. Third there are also "external locational characteristics", which influence differences in perceptions between regions (e.g., states). One may also perceive certain attributes which have "non-locational attributes". Poverty, cotton, hominy grits, and a highly stratified social system may form a class which groups the Southern States into one subjective schema. A related corollary of these observations with special ecological bearing is that two places which are geographically separated but members of a particular class may be perceived as being closer together than areas equally separated but not in the same class.

Lowenthal (1967) adds another classification scheme to the study of evaluative mapping. He divides perceptions into two dimensions : "public images", which are conscious and communicable, and "personal world", which includes more varied landscapes, imaginary places, and personal environments in which the individual is the only actor. Furthermore, " . . . in each of our personal worlds, far more than in our shared consciousness, characters of folk and fiction reside and move about some in their own lands, others in familiar places." In a more philosophical sense, "If we could not imagine the impossible, both private and public worlds would be poorer". Thus, even though perceptions are based on cognitive maps, they may form an "imagined" world which individual preferences dictate. These fantasies add to the uniqueness of individual perceptions. Beck (1967) defines this imagined world as a subjective space, made up of dreams, fantasies, spatial styles of individuals and ingrained spatial manifestations of whole cultures (also see Buttimer 1968).

The empirical test of preferences for certain areas and what they mean to the individual has been undertaken by designers, planners and a few geographers. These studies, generally employing photographs of different types of neighbourhoods, deal with the criteria of residential site selection. Flachsbart and Peterson (1973) tested the difference between preferences for certain types of neighbourhoods among poor blacks and affluent whites ; colour slides of neighbourhoods were shown to subjects. Among the significant differences found were that blacks preferred environments that possessed greater spaciousness more cultivated vegetation, newness, and smaller-scale structures. Affluent whites preferred variety, uniqueness, less cultivated vegetation, and two or three story dwellings. These preferences tend to be the opposite of conditions in which respondents live, e.g., blacks in ghettos, whites in middle-class suburbs. This result was attributed to the fact that environmental deprivation appears to inflate preferences, while environmental adaptation appears to deflate preferences.

In a series of studies of suburban preferences employing survey and related research techniques, Johnston (1971) explicitly addressed himself to the extension of the classic ecological models (especially that of Hoyt). His research on Christ Church presents an exact test of three assumptions drawn from Hoyt :

1. All people rate suburbs in the same order according to their status as living areas,
2. All people rate suburbs in the same order according to preferences for living in them, and
3. These two scales are highly correlated with suburban preference behaviour as a function of perceived socio-economic status.

In testing these assumptions, respondents were asked to rate the status ranking and relative desirability of suburbs. Status and desirability scales were compared and mental maps based on these scales were examined for sectoral bias. The author concluded :

> that sample respondents did agree on the status of various suburbs of Christ Church and on the desirability of these places as living areas . . . Apparently they assess desirability in terms of socio-economic status for the two scales were always highly correlated and to some extent these perceptions are spatially constrained by the sectors of the city in which they live.

In a panel study using university students in Australia and New Zealand (Johnston 1972), suburbs were again rated according to their social standing and their residential desirability. Results showed a "high correlation between social standing and residential desirability scales in virtually every case". In a later study in New Zealand (Johnston 1973), residents were sampled and asked to rank suburbs on five dimensions : socio-economic status, dominant lifestyle, residential desirability, physical environment, and relative location. Findings indicated that for the respondents sampled, there is a common set of dimensions along which suburbs are scaled. Yet the actual scaling of the suburbs was shown to be "a function of where the respondent lives relative to those suburbs and of the city's spatial structure".

Preference approaches

The study of preferences is similar to that of the evaluative approach, in that preferences, values, and attitudes all are assigned a role in space perception. The decisions which people make about their environments are associated with the way in which they perceive and interpret the space that surrounds them. The knowledge they receive through diverse channels of information also plays a part. These types of inputs become more important when the decision to migrate arises during the life-cycle of a household. At this point, images of potential new locales gained through travel and other day-to-day experiences within the city are organized.

According to Adams (1969) urban residents accumulate information and create sets of ideas through their experiences ;

but the information accumulated seldom increases greatly beyond a certain threshold no matter how long one lives in an area. Adams calls this the "locational awareness" or "spatial vocabulary" which is carried around by the potential migrant and which is presumed to be sectorally shaped. (see Donaldson 1973, Donaldson and Johnston 1973, Johnston 1971). Thus there is always a "boundary" which affects the movement of urban residents. Adams found that ghetto dwellers for example had a sharp image of their surrounding area, but that this image became blurred toward the outer boundaries of their home location. The image of affluent whites was also restricted but in a different way, e.g., their distant images were limited also but they tended to have more images scattered around the city. Wolpert (1965) formulated a "threshold of utility" by which the resident measures his present location against alternate locations. The evaluation of alternatives takes the form of a search behaviour in which the alternative locations are those areas that the potential migrant is most familiar with. This new physical location (of the household) is viewed as being within an action space that the migrant has the most information about and "... the subjective utility or preferences he associates with this location" (Horton and Reynolds 1971).

The role of preferences in the migration decision has been discussed by Brown and Moore (1970). When disruption of normal behaviour—(stressors) such as a new job or higher status—brings a household to the decision to migrate or not to migrate, upper and lower limits of criteria for a new residence are formulated. This is an evaluative process by which the migrant eliminates areas that do not meet the criteria before the search begins. Through diverse channels of information (friends, other migrants, mass media) and travel experiences, the potential migrant searches familiar areas. In general, the migrant forms a subjective evaluation of potential places that are within an area around his home sector, while the awareness space is bounded by unfamiliar areas that are not in his "activity space". Following upon this work with a more sociological orientation, White (1974) measured the space utility of potential residences finding that preferences for residential areas vary by group (age cohort and ethnic group). He also found that the best predictor of attraction for an area is "... strongly related to an aggregate residential preference value attributed to that city by potential migrants (i.e., group norms)". Clark and Cadwallader (1973) used census block data to map the preferences of residents (by income and ethnic categories) of the Los Angeles area. The respondents were asked which area they would most like to live in, keeping in mind their total family income (thus hoping to control for fantasies). A majority of the respondents chose the community they were already living in or one close to it. Lower income groups (keeping in mind they were aware of the income constraints) preferred areas located around the CBD, middle income groups preferred areas outside the central area, while high income groups had the simplest and most restricted preference pattern with the least number of preferred areas. Ethnic groups like Blacks and Mexican-Americans preferred areas with the highest concentration of members of their own groups. But overall, the differences between all the groups, income or ethnic group, were found to be small.

Summary of geographic research

This catalogue of geographic research reveals four points to be stressed in summary. First, in this literature there is an exploration into the many ways in which urban space is bound by subjectivity. In individual cognition, evaluation (of self and others) and spatial preferences consciousness is seen to impose structure on the "given" environment. Second, most of these explorations have been pursued from a behavioural perspective (in contrast to the community and urban politics perspectives familiar in sociology). As a result, an extensive set of survey findings has been accumulated but which awaits theoretical synthesis. Third, the principal non-spatial variables included in the geographic research designs tend to be psychological rather than sociological or economic. Although samples tend to be small and somewhat unsystematic in this research, generalizations tend to be made to "people", "actors", or "migrants". Fourth, when sociological categories are employed (as they are in the work of Orleans, Petersen and Johnston) the results tend to support the classic models. While conclusions drawn from these few studies must be highly qualified, there does not appear to be a great discrepancy between the city as the classic ecologists viewed it and the city which the occupants see.

The first two points suggest that geographers have explored areas which have, in general, been overlooked in sociology. There are many ways, both substantive and methodological, in which future ecological research can benefit from the complementarity of the two disciplines as they focus on the role of subjectivity. The third point, similarly, suggests that the perception-psychology perspective many have been overlooked by sociologists. But it also suggests a weakness in the geographic literature which might be addressed by developing a stronger role for sociology in the education of geographers. The fourth point is of interest in the light of our attempt to understand the relative lack of development of the subjective approach in sociology, and it also may have something to say about the value of comparative ecological research.

Subjective ecology was formulated explicitly to explain the distortions in the classic models and rationalist formulations. To achieve this, a new set of factors—the subjective component—was introduced into the explanation of land use and social distributions. At the same time, and to this day, objectivity-oriented ecologists have demonstrated that the classic models do fit the social geography of cities (in industrialized countries) *well enough* (i.e., within tolerable statistical variance). The (albeit modest) research which has been conducted on the correspondence between mental maps and the classic models shows that the two are not incompatible.

The lack of development of the subjective approach within mainstream urban ecological studies may be due, in part, to a basic correspondence between the subjectivity implicit in the classic models and the subjectivity of actors. Because both reflects the same culture milieu, the contribution of subjective components to the reduction of variance in applications of the classic models (especially when the cost of gathering subjective data is taken into account) is negligible. While knowledge of values, interests and symbolic process may enrich the understanding of ecological process and structure, they are already well enough incorporated into the objective models to make them unnecessary. In brief, the distortions introduced by considering only census, economic and transport data in the study of urban ecology (in industrialized countries) have not been great enough to justify the subjective approach ; not because the subjective approach fails to focus on an important set of factors, but because the implicit subjective assumptions of the models adequately reflect aggregate sentiments, values and cognition.

If this last observation is correct, then the subjective approach might be shown to make a difference in a culture in which the subjective assumptions of the models do not correspond to those of urban dwellers. In my research in comparative urban ecology, I have found that the classic models are put to a more challenging test in cultures in which Western values, interests and related cultural elements (assumed by the models) are not widely held. In the following section, this point is illustrated in an examination of aspects of the ecological structure of Indian cities.

AN APPLICATION OF SOME SUBJECTIVE CONCEPTS IN TWO INDIAN CITIES

Studies of Madras and Ahmedabad (see Chapters 7 and 8) were undertaken with the object of examining the degree to which the ecological structures of these cities are modernizing. In order to avoid some of the problems of validity, the research design of these studies incorporated four elements which had not been used in earlier work :

(1) Time series comparisons of census data and historical material (from various secondary sources) were integrated with the structure analysis.

(2) Data were gathered and aggregated for various levels, ward, division, block, in order to avoid artificialities inherent in large tracts.

(3) A social survey of approximately 2,000 households was conducted (in each city). The interview scheduled contained three types of questions :

(a) "demographic", standard social area variables not included in the census of India,
(b) "social structural", detailed information on family structure, employment and caste, and
(c) "subjective", information on respondents views of the city and self.

(4) A social anthropological perspective (based on Wiebe 1976) was provided through the use of informants and participant observation techniques.

While some of the findings of these studies have been reported in Chapters 7 and 8 of this volume, much additional analysis and interpretation also remains to be done. Nevertheless, some provisional conclusions can be offered :

(1) The classic ecological models do not fit the actual structures of these cities (Weinstein 1976 a, b).
(2) In particular, the segregation of groups and land use types assumed by the model is too broad and too regular to capture the actual—smaller scale—segregation which exists (Weinstein 1974).
(3) These cities cannot be said to be modernizing in a cumulative sense. A more accurate way to depict current social change in Madras and Ahmedabad is as an increasingly complex inter-penetration of modern and traditional/local forms (Weinstein and McNulty 1979).
(4) There is an aspect of the ecological structure of these cities which is comprehensible only when knowledge of specifically cultural (e.g., religion, caste) or other subjective components is considered.

In light of our discussion of the subjective approach in sociology and geography, the fourth point appears to be central in explaining the lack of correspondence between model and observations indicated in the first three points. That is, one might hypothesize that the reason why the classic models do not apply very well in Indian cities is that the models assume a shared subjectivity, a culture, which is not present. Similarly, the lack of a tendency toward modernization—where one should be manifest—might be attributed to subjective factors which (intentionally or not) offer resistance.

In Madras, spatial juxtapositions of groups and land uses which would seem anomalous from the perspective of the model metropolis are the rule. The city displays a very small scale at which land use and group segregation are organized. While one can speak broadly of middle or working class neighbourhoods, areas of Muslim concentration, and so on, considerable variation exists within census tracts. Distinctions between residential, commercial, recreational and even industrial areas of the city are, at best, statements of general tendencies. For within a few blocks of nearly any point in the city these and other land uses can be found.

Groups which are far apart in relation to social hierarchies (e.g., caste) can often live very close together (while still observing strict codes of social segregation) : servants living near prominent families, houseless persons sleeping near middle class apartments, slum dwellers building shanties next to expensive commercial or residential property. In sum, the ecological structure of Madras is organized on a scale principle quite different from that assumed by objective models.

The classic models of ecological structure, research on evaluative mental mapping and earlier, census-based, studies of Indian cities all would suggest that a single hierarchy of neighbourhood desirability guides spatial behaviour in Madras. This research on social and ecological structure there suggests otherwise. The desirability of a neighbourhood tends not to depend on its location relative to the central business district or other setting. Instead, controlling for self selection (which in Madras, as elsewhere, is very frequent), the desirability of an area to a resident is positively related to the similarity of the area's social characteristics to those of the rater. In Parsonian terms, particularistic criteria of neighbourhood preference exist in Madras, yet the model is premised on universalistic criteria.

The Ahmedabad study suggest that, as in Madras, the objective ecological approach leaves unexplained much of the essential character of the city's sociogeographic systems. While the classic models anticipate regular and segregated patterns of land use, the city is organized on different principles. Ahmedabad, like Madras, has been shaped by modernizing forces ; and these forces have had their effect on contemporary ecological structure. Yet, it is a city whose history, traditions and ethnic conflicts have also had great impact. The ecological structure of the city—as reflected in its mills, walls and slums—is the outcome of the interaction among all of these factors. Thus the land use and population distributions are far more complex and irregular than those associated with the model modern metropolis.

Madras and Ahmedabad represent two very different types of cities. One is a colonial city in the south-eastern Tamil-speaking section of India. The other is a Muslim city of the fifteenth century in the north-west, Gujarat-Bombay region. One city is a cosmopolitan sea-port with a commercial-administrative organization, the other is an inland (and insular) industrial centre. These differences are reflected in many ways in the ecological structures of these cities.

Both cities, however, exhibit a similar underlying complexity which is probably shared with other cities in India. To varying degrees, each city contains a strong local/traditional orientation which runs counter to modern/"rational" norms and interests. As a result, the reality of land use and population distribution is multi-leveled. Traditional codes of hierarchy, segregation and interaction prevail ; multiple land use is the rule. In no instance are these elements unaffected by the historical fact of colonial rule. Yet colonialism and modernization have had to make accommodations to the persistent influence of local culture.

When the Chicago School sociologists characterized urban life in terms of its density, complexity and heterogeneity, their principal cases were American and European. The development of models and theories of urban ecological structure based upon the work of the Chicago School extended these observations in providing an increasingly organized (and a more empirically grounded) sense of these characteristics. In cities like Ahmedabad and Madras, which in addition to the familiar sources of density, complexity and heterogeneity, have also been influenced by colonial and imperial relations, conquest, and conscious modernization policies, the classic models and theories prove to be excessively simple (or inadequate in other ways). The order of density, complexity, and heterogeneity of these cities far surpasses that anticipated by existing models. In most respects, these are the most profoundly urban of any of the world's large cities.

This brief review of some of the findings on Madras and Ahmedabad cannot demonstrate in any systematic way *how* subjective components, unanticipated by classical ecological approaches, affect land use in Indian cities. But it should be clear enough that it is the subjectivity of *Madrasis* and *Ahmedabadis* which has shaped their cities. The examination of subjective approaches in ecology and geography points to a rich source of theory, hypotheses and research techniques with which to pursue future research.

CONCLUSION

Local cultural views, ethnic and caste loyalties, traditional segregation principles, particularistic cannons of neighbourhood desirability, and norms of intensive land use appear to be among the most prominent subjective elements influencing ecological structure in Indian cities. The details of these relationships remain obscure. Yet, to have identified their role against the background of subjective ecology and mental mapping is to have taken an important first step toward providing a sociological account of Indian urban ecological structure. The next several steps involve close operationalization of subjective concepts and the direct testing of hypotheses regarding, e.g., the role of segregation especially with comparative data. To aid in this project, a virtually unexploited body of literature awaits application.

The most appropriate sociological metaphor for urban ecological structure comes neither from systems theory nor from social geography. Rather, it should be characterized as a "collective representation" (Weinstein 1974). That is, a city's distribution of land uses and populations is conceived of as the, partly-intended, product of social behaviour which "stretches not only into space but into time as well" (Durkheim 1954). As a collective representation, it incorporates the image of its creator, that is, urban society. The power of the subjective approach in ecological studies lies in its focus on those aspects of urban social life—values, interests and related cultural products—which are most likely to leave their stamp on ecological structure. In attending to the subjective component, one is using ecology not as an end in itself but rather as an aid to insight on social behaviour. In the absence of a subjective focus, social behaviour is, as Firey so forcefully argued, always viewed as a dependent variable, a passive element.

Firey, too, has noted the appropriateness of Durkheim's views to the study of contemporary ecological structure. He speaks, for example, of the principle :

> which Durkheim saw when he observed that men's moral conscience could not long endure without the aid of external reminders. The "collective conscience" goes through continual variations in intensity so that it would be liable to extinction during its weaker phases unless there were something more enduring to summon it back into being (Firey 1947).

One "construction" which serves this function is urban ecological structure. Firey and others have extended this metaphor in their stress on the subjective component ; yet this has not been the prevailing image of ecological structure in most sociological studies.

BIBLIOGRAPHY

Abrams, C. ; (1964) *Man's Struggle for Shelter in an Urbanizing World*, Cambridge, MIT Press.

Abu Lughod, J. ; (1968a) "The City is Dead—Long Live the City : Some Thoughts on Urbanity", in S. Fava (ed), *Urbanism in World Perspective*, Illinois, Crowell Company.

——; (1968b) "The Factorial Ecology of Cairo", unpublished paper, Department of Sociology, Northwestern University.

——; (1969) "Testing the Theory of Social Area Analysis : The Ecology of Cairo, Egypt", *American Sociological Review*, 34(2), pp. 198-212.

Adams, J.S. ; (1969) "Directional Bias in Intra-Urban Migration", *Economic Geography*, 45, pp. 302-23.

Alihan, M.A. ; (1938) *Social Ecology : A Critical Analysis*, Columbia, Columbia University Press.

Anderson, T. and J. Bean ; (1961) "The Shevky—Bell Social Areas : Confirmation of Results and Reinterpretation", *Social Forces*, 40(2), pp. 119-24.

Anderson, T. and J. Egeland ; (1961) "Spatial Aspects of Social Area Analysis", *American Sociological Review*, 26(3), pp. 392-98.

Ayyar, C.P.V. ; (1916) *Town Planning in Ancient Dekkan*, Madras, Law Publishing House.

Beck, R. ; (1967) "Spatial Meaning and the Properties of the Environment", in D. Lowenthal (ed.), *Environmental Perception and Behaviour*, Research Paper No. 109, Department of Geography, Chicago, University of Chicago.

Bell, W. ; (1953) "The Social Areas of the San Francisco Bay Region", *American Sociological Review*, 18, pp. 29-47.

——; (1955) "Economic Family and Ethnic Status : An Empirical Test", *American Sociological Review*, 22, pp. 45-52.

——; (1969) "Urban Neighbourhoods and Individual Behaviour", in (eds) P. Meadows and E. Mizruchi, *Urbanism,*

Urbanization and Change, Reading, Massachusetts, Addison-Wesley.

Berry, B.J.L. ; (1965) "Internal Structure of the City", *Law and Contemporary Problems*, 30(1), pp. 111-19.

———; (1966) *Essays on Commodity Flows and the Spatial Structure of the Indian Economy*, Research Paper No. 111, Department of Geography, Chicago, University of Chicago.

———; (1971) "The Passing of Central Place Theory, Research Institute—Lectures on Geography", *Special Report*, Mimeographed, pp. 113-18.

Berry, B.J.L. and J. Kasarda ; (1977) *Contemporary Urban Ecology*, London, MacMillan.

Berry, B.J.L. and P.H. Rees ; (1969) "The Factorial Ecology of Calcutta", *American Journal of Sociology*, 74, pp. 445-91.

Berry, B.J.L., J.W. Simmons and R.J. Tennant ; (1963) "Urban Population Densities : Structure and Change", *Geographical Review*, 50, pp. 389-405.

Berry, B.J.L. and H. Spodek ; (1971) "Comparative Ecologies of Large Indian Cities", *Economic Geography*, 47, pp. 266-85.

Beshers, J.A. ; (1962) *Urban Social Structure*, Glencoe, The Free Press.

Bopegamage, A. ; (1966) "A Methodological Problem in Indian Urban Sociological Research", *Sociology and Social Research*, 50, pp. 236-40.

Bose, N.K. ; (1965) "Calcutta : A Premature Metropolis", *Scientific American*, 213 (3), pp. 90-102.

———; (1968) *Calcutta 1964 : A Social Survey*, Calcutta, Lalvani Publishing House.

Breese, G. ; (1966) *Urbanization in Newly Developing Countries*, New York, Prentice Hall.

Brown, L.A. and E.G. Moore ; (1970) "The Intra-Urban Migration Process : A Perspective" in L.S. Bourne (Ed.), *Internal Structure of the City*, Oxford, Oxford University Press.

Brush, J.E. ; (1962) "The Morphology of Indian Cities", in R. Turner (ed.), *Indian's Urban Future*, pp. 57-70, Berkeley and Los Angeles, University of California Press.

———; (1968) "Spatial Patterns of Population in Indian Cities", *Geographical Review*, 58, pp. 362-91.

Brush, J.E. ; (1970) "Some Dimensions of Urban Population Pressure in India", in W. Zelinsky, L.A. Kosinki and A.M. Prothero (eds.), *Geography and a Crowding World*, New York and London, Oxford University Press.

———; (1973) "Application of a Model to the Analysis of Population Distribution in Indian Cities", in R.L. Singh (ed.), *Urban Geography in Developing Countries*, pp. 28-49, Varanasi, National Geographical Society of India.

———; (1976) "Calcutta" *Focus* (American Geographical Society), 27 (1), pp. 1-8.

Burgess, E.W. ; (1925) "The Growth of the City : An Introduction to a Research Project", in R. Park, E.W. Burgess and R.D. McKenzie (eds.), *The City*, pp. 47-62, Chicago, University of Chicago Press.

———; (1928) "Residential Segregation in American Cities", *Annals of the American Academy of Political and Social Science*, 140, p. 108.

Buttimer, A. ; (1968) "Social Space in an Interdisciplinary Perspective", *The Geographical Review*, 58, pp. 417-26.

Caplow, T. ; (1949) "The Social Ecology of Guatemala City", *Social Forces* 28, pp. 113-33.

Chakraborty, S. ; 1978 "Urban Economic Base—A Review of Contemporary India Experience", Unpublished Paper, Indo-French Seminar, New Delhi.

Chatterjee, A.B. ; (1960) "Howrah : An Urban Study", Unpublished Ph.D. dissertation, University of London, London.

Chombart de Lauwe, P.H. ; (1952) *Paris et l' agglomeration Parisienne*, Paris, Presses Universitaires de France.

Choskey, R.B. ; (1968) *Economic Life in Bombay and Gujarat*, New York.

Clark, W.A.V. ; (1969) "Information Flows and Intra-Urban Migration : An Empirical Analysis", *Proceedings of the Association of American Geographers*, 1, pp. 38-42.

Clark, W.A.V. and M. Cadwallader ; (1973) "Residential Preferences : An Alternate View of Inter-Urban Space", *Environment and Planning*, 5, pp. 693-703.

Cox, O. ; (1969) "The Preindustrial City Reconsidered", in P. Meadows and E. Mizruchi (eds.), *Urbanism, Urbanisation and Change*, Massachusetts, Addison-Wesley.

Cox, K. and G. Zannares ; (1973) "Designative Perception of Macro-Spaces : Concepts, a Methodology and Application", in R.M. Downs and D. Stea (eds.), *Image and Environment*, United Kingdom, Aldine Publishing Company.

Davis, A., B.B. Gardner, and M.R. Gardner ; (1941) *Deep South*, Chicago, University of Chicago Press.

Deshpande, C.D. ; (1973) "The suburbs of Greater Bombay", unpublished paper, Centre for Urban Research, Department of Geography, Osmania University, Hyderabad.

Dixit, K.R. ; (1959) "The Problems of land use in the Planning of Greater Bombay", *Bombay Geographical Magazine*, 6-7, pp. 65-79.

Donaldson, B. ; (1973) "An Empirical Investigation into the Concept of Sectoral Bias in the Mental Maps, Search Spaces and Migration Pattern of Intra-Urban Migration", *Geografiska Annaler*, 55, pp. 13-33.

Donaldson, B. and R.J. Johnston ; (1973) "Intra-Urban Sectoral Mental Maps : Further Evidence from an Extended Methodology", *Geographical Analysis*, 5, pp. 45-54.

Dotson, F. and L.O. Dotson ; (1954) "Ecological Trends in the City of Guadalajara, Mexico", *Social Forces*, 32, pp. 367-74.

Downs, R.M. ; (1970) "Geographical Space Perception : Past Approaches and Future Prospects", in C. Board (ed.), *Progress in Geography Volume 2*, London, Edward Arnold Ltd.

Downs, R.M. and D. Stea ; (1973) "Cognitive Maps and Spatial Behaviour : Process and Products", in R.M. Downs and D. Stea (eds.), *Image and Environment*, United Kingdom, Aldine Publishing Company.

Duncan, O.D. ; (1955) Untitled review of Social Area Analysis by E. Shevky and W. Bell, American Journal of Sociology, 61, p. 84.

———; (1959) "Human Ecology and Population Studies", in P.M. Hauser and O.D. Duncan (eds.), *The Study of Population*, pp. 678-716, Chicago, University of Chicago Press.

Dupuis, J. ; (1960) *Madras et Le Nord du Coromandel*, Paris, Librarie d' Amerique et d' Orient.

Durkheim, E. ; (1954) *The Elementary Forms of the Religious*

Life, Glencoe, The Free Press.

Dutt, B.B. ; (1925) *Town Planning in Ancient India*, Culcutta, Thacker Pink & Co.

Epstein, T.S. ; (1962) *Economic Development and Social Change in South India*, Manchester, Manchester University Press.

Firey, W. ; (1945) "Sentiment and Symbolism as Ecological Variables", *American Sociological Review*, 10, pp. 140-48.

———; (1947) *Land Use in Central Boston*, Cambridge, Harvard University Press.

Form, W. ; (1954) "The Place of Social Structure in the Determination of Land Use : Some Implications for a Theory of Urban Ecology", *Social Forces*, 32, 317-23.

Flachsbart, P.G. and G.L. Petersen ; (1973) "Dynamics of Preference for Visual Attributes of Housing Environments", in W.F.E. Preiser (ed.), *Environmental Design Research Volume 1*, Dowden, Hutchinson and Ross.

Fox, R.G. ; (1969) *Tezibazar*, Ithaca, Cornell University Press.

Friedmann, R.P. and J. Miller ; (1965) "The Urban Field", *Journal of the American Institute of Planners*, 31, pp. 312-19.

Gadgil, D.R. ; (1945) *Poona : A Socio-economic Survey Part 1*, Pune, Gokhale Institute of Politics and Economics.

———; (1952) *Poona : A Socio-economic Survey Part 2*, Pune, Gokhale Institute of Politics and Economics.

———; (1965) *Sholapur City : Socio-economic Studies*, Pune, Gokhale Institute of Politics and Economics.

Gans, H. ; (1962) *The Urban Villagers*, New York, The Free Press.

———; (1967) *The Levittowners*, New York, Pantheon Books.

Gettys, W.E. ; (1940) "Human Ecology and Social Theory", *Social Forces*, 18, pp. 409-476.

Ghosh, A. ; (1961) "Calcutta : The Primate City", Monograph, Census of India.

Ghurye, C.S. ; (1961) *Caste, Class and Occupation*, Bombay, Popular Book Depot.

Gillian, K.L. ; (1968) *Ahmedabad*, Berkeley, University of California Press.

Gist, N.P. and S. Fava ; (1974) *Urban Society*, Illinois, Crowell Company.

Golledge, R.G., R. Briggs and V. Demko. ; (1969) "The Configuration of Distance in Intra-Urban Space", *Proceeding of the Association of American Geographers*, 1, pp. 60-65.

Gould, P. ; (1967) "Structuring Information on Spatio-Temporal Preferences", *Journal of Regional Science*, 7, pp. 2-16.

——— ; (1968) "Problems of Space Preference Measures and Relationships", *Geographical Analysis*, 1, pp. 31-44.

——— ; (1969) "Structure of Space Preferences in Tanzania", *Area*, 4, pp. 29-35.

Gould, P. and R. White ; (1974) *Mental Maps*, New York, Penguin Books.

Gower, J.C. ; (1966) "Some Distance Properties of Latent Root and Vector Methods used in Multivariate Analysis", *Biometrika*, 53, pp. 325-28.

Gusfield, J. ; (1967) "Tradition and Modernity : Misplaced Polarities in the Study of Social Change", *American Journal of Sociology* 72(4).

Hanspal, R. ; (1977) "Urban Structure of Bangalore City : A Spatial Analysis", Unpublished Ph.D. dissertation, Institute for Social and Economic Change, Bangalore.

Hasan, M.F. ; (1970) *Bangalore Through the Centuries*, Bangalore, Historical Publishers.

Harris, C.D. and E. Ullman ; (1945) "The Nature of Cities", *Annals of the American Academy of Political and Social Science*, 242, pp. 7-17.

Hawley, A. ; (1950) *Human Ecology*, New York, Ronald Press.

——— ; (1968) "Human Ecology", *The International Encyclopedia of the Social Sciences*, 4, pp. 328-37.

Hawley, A. and O.D. Duncan ; (1957) "Social Area Analysis : A Critical Appraisal", *Land Economics*, 33, pp. 337-45.

Hayner, N.S. ; (1945) "Mexico City : Its Growth and Configuration", *American Journal of Sociology*, 50, pp. 295-304.

Hollingshead, A.B. ; (1939) "Human Ecology", in R. Park (ed.), *An Outline of the Principles of Sociology*, New York.

——— ; (1947) "A Re-examination of Ecological Theory", *Sociology and Social Research*, 31, pp. 194-204.

Horan, P. ; (1978) "Its Status Attainment Research Atheoritical ?" *American Sociological Review*, 43.

Horton, F.E. and D. Reynolds ; (1971) Effects of Urban

Spatial Structure on Individual Behaviour", *Economic Geography*, 47, pp. 36-48.

Hoyt, H. ; (1933) *One Hundred Years of Land Values in Chicago*, Chicago, University of Chicago Press.

——— ; (1939) *The Structure and Growth of Residential Neighbourhoods in American Cities*, Washington, Federal Housing Administration.

Johnston, R.J. ; (1971) "Mental Maps of the City : Suburban Preface Patterns", *Environment and Planning*, 3, pp. 63-72.

——— ; (1972) "Activity Spaces and Residential Preferences : Some Tests of the Hypothesis Sectoral Mental Maps", *Economic Geography*, 8, pp. 199-210.

——— ; (1973) "Spatial Patterns in Suburban Evaluations", Unpublished paper, Department of Geography, University of Canterbury, New Zealand.

Jonassen, C. ; (1954) "Cultural Variables in the Ecology of an Ethnic Group", *American Sociological Review*, 19.

Karve, D.D. ; (1963) *The New Bramhans : Five Maharashtrian Families*, Berkeley, University of California Press.

Kirschenbaum, A. ; (1969) "Flight from Suburbia : A Demographic Analysis", in P. Meadows and E. Mizruchi (eds.), *Urbanism, Urbanisation and Change*, Reading, Massachusetts, Addison-Wesley.

Kohl, J.G. ; (1841) *Der Verkehr und die Ansiedlung der Menschen in ihrer Abhangigkeit Von der Gestaltung der Erdoberflache*, Leipzig, Arnoldische Buchhandlung.

Kosambi, M. ; (1980) "Bombay and Poona—A Sociological Study of Two Indian Cities 1650-1900", Ph.D. dissertation, Department of Sociology, University of Stockholm, Sweden.

Latif, A. ; (1974) "Factor Structure and Change : Analysis of Alexandria, Egypt, 1947-1960", in K.P. Schwirian (ed.), *Comparative Urban Structure*, Boston, D.C. Heath.

Lowenthal, D. ; (1967) *Environmental Perception and Behaviour*, Research Paper No. 109, Department of Geography, Chicago, University of Chicago.

Lynch, K. ; (1960) *The Image of the City*, Massachusetts, MIT Press.

Malik, R.A. ; (1965) "Cities of the Upper Indo-Gangetic Plain", *Pakistan Geographical Journal*, 10, pp. 61-72.

Mann, P.H. ; (1965) *An Approach to Urban Sociology*, London,

Routledge and Kenganpaul.

Mayer, A.C. ; (1960) *Caste and Kinship in Central India : A Village and its Origin*, Berkeley, University of California Press.

McGee ; (1971) *The Urbanization Process in the Third World*, London, Bell & Sons.

McKenzie, R.D. ; (1933) "Human Ecology", *Encyclopedia of the Social Sciences*, New York, McMillan & Co.

Mehta, S.K. ; (1968) "Patterns of Residence in Poona (India) by Income, Education and Occupation (1937-1965)", *American Journal of Sociology*, 73(4), pp. 496-508.

——— ; (1969) "Patterns of Residence in Poona, India, by Caste and Religion : 1822-1965", *Demography*, 6(4), pp. 473-91.

Michelson, W. ; (1971) "An Empirical Analysis of Urban Environmental Preferences", in L.S. Bourne, (ed.), *Internal Structure of the City*, Oxford, Oxford University Press.

Mitra, A. ; (1963) *Calcutta : India's City*, Calcutta, New Age Publishers Private Ltd.

Murvar, V. ; (1966) "Some Tentative Modifications of Weber's Typology : Occidental Versus Oriental Cities", *Social Forces*, 44, pp. 381-89.

Nissel, H. ; (1977) "Bombay—Untersuchungen Zur Structure and Dynamik einer indischen Metropole", *Berliner Geographische Studien*, Band 1, Berlin, Institut Fur Geographie der Technischen Universitat Berlin.

Orleans, P. ; (1973) "Differential Cognition of Urban Residents : Effects of Social Scale on Mapping", in R.M. Downs and D. Stea (eds.), *Image and Environment*, United Kingdom, Aldine Publishing Company.

Pampel, F. and H. Choldin ; (1978) "Urban Location and Segregation of the Aged", *Social Forces*, 56(4).

Park, R.E. ; (1916) "The City : Suggestions for the Investigation of Human Behaviour in the Urban Environment", *The American Journal of Sociology*, 20, pp. 577-612.

——— ; (1936) "Human Ecology", *American Journal of Sociology*, 42, pp. 1-15.

Park, R.E., E.W. Burgess and R.D. McKenzie (eds.) ; (1925) *The City*, Chicago, University of Chicago Press.

Parvathi, C. ; (1977) "Market Centres and Spatial Development in Coimbatore Nilgiris Region", Unpublished Ph.D. dissertation, Mysore University, Mysore.

Patel, S.B. ; (1974) "The Future of Bombay—An Alternative Approach to Urban Planning", in A. Bose, P.B. Desai, A. Mitra and J.N. Sharma, (eds.), *Population in India's Development 1947-2000*, pp. 306-316, Delhi, Vikas Publication House.

Pedersen, P.O. ; (1967) *Modeller for Befolknings struktur og Befolknings-udrikling : Storby morader Specielt med Henblik pa Storkobenharn*, Copenhagen, State Urban Planning Institute.

Prakasa Rao, V.L.S. and R. Ramachandran ; (1971) "Mobility and Choice of Central Places", *Analytical Geography*, 1.

Prakasa Rao, V.L.S. and V.K. Tewari ; (1979) *The Structure of an Indian Metropolis : A Study of Bangalore*, New Delhi, Allied Publishers.

Quinn, J.A. ; (1940) "The Burgess Zonal Hypothesis and its Critics", *American Sociological Review*, 5, pp. 210-18.

Rajagopalan, C. ; (1963) "An Ecological Analysis of the Growth of Bombay City", *Geografia*, 2, pp. 99-105.

Ramu, G.N. and P. Wiebe ; (1970) "Elections in Peddur", *Eastern Anthropologist*.

Rao, H. ; (1930) *Mysore Gazetteer*, 5, Bangalore, Government Press.

Raza, M. ; (1979) "Spatial Organization and Urbanization in India : A Case Study of Underemployment", in R.P. Misra and K.V. Sudaram (eds.), *Rural Area Development Perspectives* and Approaches, Delhi, Sterling, pp. 333-77.

Rees, P.H. ; (1969) "The Factorial Ecology of Metropolitan Chicago", in B.J.L. Berry and F.E. Horton (eds.), *Geographic Perspectives on Urban Systems*, pp. 276-397, Englewood Cliffs, Prentice Hall.

——— ; (1971) "Factorial Ecology : An Extended Definition, Survey and Critique of the Field", *Economic Geography*, 47(2), pp. 220-33.

Schwirian, K. ; (1971) "Some Analytical Problems in the Comparative Test of Ecological Theories", Unpublished paper, Ohio State University, Ohio.

——— ; (1974) (ed.), *Comparative Urban Structure*, Boston, D.C. Heath.

Schwirian, K. and M. Matre ; (1974) "The Ecological Structure of Canadian Cities", in K. Schwirian (ed.), *Comparative Urban Structure*, Boston, D.C. Health.

Shah, R. ; (1976) "An Enquiry into Housing Conditions of Industrial Workers in Ahmedabad", Unpublished M. Arch. Thesis, School of Planning, Ahmedabad.

Shevky, E. and W. Bell ; (1955) *Social Area Analysis : Theory, Illustrative Application and Computational Procedures*, Palo Alto, California.

Shevky, E. and M. Williams ; (1949) *The Social Areas of Los Angeles*, Berkeley, The University of California Press.

Simmel, G. ; (1923) *Sociologie*, Munchen, Duncker and Humblot.

Singh, H. ; (1975) "Patterns of Socio-Economic Structure in Indian Cities", *Asian Profile*, 3, pp. 283-98.

Sion, J. ; (1909) "Les Paysans de la Normandie Orientale : Pays de Caux, Bray, Vexin Normand, Vallee de la Seine", *Etude Geographique*.

Sjoberg, G. ; (1960) *The Pre-Industrial City, Past and Present*, New York, The Free Press.

Smailes, A.E. ; (1953) *The Geography of Towns*.

Sonnenfeld, J. ; (1968) "Geography, Perception and the Behavioural Environment", in P.W. English and R.C. Mayfield (eds.), *Man, Space and Environment*, Oxford, Oxford University Press.

Sorokin, P.A. ; (1943) *Sociocultural Casuality, Space and Time*, Durham, Duke University Press.

Spate, O.H.K. and E. Ahmad ; (1950) "Five Cities of the Indo-Gangetic Plain", *American Geographical Review*, 40, pp. 260-78.

Spate, O.H.K. and A.T.A. Learmonth ; (1967) *India and Pakistan*, England, Methuen.

Spodek, H. ; (1965) "The Manchesterization of Ahmedabad", *Economic and Political Weekly*, March 13.

———; (1971) "On the Origin's of Gandhi's Political Methodology", *Journal of Asian Studies*, 30.

Stea, D. ; (1967) "Reasons for our Moving", *Landscape*, 17, pp. 27-28.

———; 1969 "The Measurement of Mental Maps : An Experimental Model for Studying Conceptual Space", in K. Cox and R. Gooledge (eds.), *Behavioural Problems in Geography : A Symposium*, pp. 228-253, Illinois, North-western University Press.

Suttles, G.D. ; (1968) *The Social Order of the Slum*, Chicago, University of Chicago Press.

———; (1972) *The Social Construction of Communities*, Chicago, University of Chicago Press.

Sweetser, F.L. ; (1965a) "Factorial Ecology : Helsinki, 1960", *Demography*, 2, pp. 372-85

———; (1965b) "Factor Structure as Ecological Structure in Helsinki and Boston", *Acta Sociologica*, 26, pp. 205-25.

Trowbridge, C.C. ; (1913) "Fundamental Methods of Orientation and Imaginary Maps", *Science*, 38, pp. 888-97.

Tyree, A. and J. Hodge. ; (1978) "Editorial Forward : Five Empirical Landmarks", *Social Forces*, 56(3).

Van Arsdol, M., S. Cammilleri and C. Schmid. ; (1958) "The Generality of Urban Social Area Indices", *American Sociological Review*, 23(3).

Wallerstein, J. ; (1974) *The Modern World System*, New York.

Ward, J.H. ; (1963) "Hierarchial Grouping to Optimise an Objective Function", *Journal of American Statistical Association*, 58(1), pp. 236-43.

Weber, A. ; (1929) Uber den Standort der Industrien 1 : Reine Theorie des Standorts. Tubingen : Mohr. Translated by C.J. Friedrich, *Theory of the Location of Industries*, Chicago, University of Chicago Press.

———; (1958) *The City*, Glencoe, The Free Press.

Weinstein, J. ; (1974) *Madras : An Analysis of Urban Ecological Structure in India*, Beverly Hills, Sage Publications.

———; (1975) "Review of K. Schwirian, Comparative Urban Structure", in *Contemporary Sociology*.

———; (1976a) "Madras : Theoretical, Technical and Empirical Issues", in G.R. Gupta (ed.), *Main Currents in Indian Sociology*, Delhi, Vikas Press.

———; (1976b) "Modernization and the Convergence of Urban Ecological Structure in India", in V.L.S. Prakasa Rao (ed.), *Indian Geographical Society Golden Jubilee Volume*, Madras, Indian Geographical Society.

———; (1976c) "Comparative Urban Studies : A Crisis in the

Urban Paradigm", *Indian Journal of Comparative Sociology*, 2(2).

Weinstein, J. and M. McNulty ; (1979) "The Interpretation of Modern and Traditional Structures : A Spatial Perspective", *Studies in Comparative International Development*, 14(2).

Wheatley, P. ; (1967) "Proleptic Observations on the Origin of Urbanism", in (eds.), R.W. Steel and R. Lawton, *Liverpool Essays in Geography*, pp. 315-45.

White, S.E. ; (1974) "Residential Preferences and Urban In-Migration", *Proceedings of the Association of American Geographers*, 6, pp. 47-50.

Whyte, W. ; (1943) *Street Corner Society*, Chicago, University of Chicago Press.

Wiebe, P.D. ; (1976) *Social Life in an Indian Slum*, Delhi, Vikas Publishing Company.

Willhelm, S.M. ; (1962) *Urban Zoning and Land Use Theory*, New York, The Free Press.

———; (1964) "The Concept of the 'Ecological Complex' : A Critique", *American Journal of Economics and Sociology*, 23(3), pp. 241-48.

Wolpert, J. ; (1965) "Behavioural Aspects of the Decision to Migrate", *Papers and Proceedings of the Regional Science Association*, 15, pp. 159-69.

Zorbaugh, H.N. ; (1926) "The Natural Areas of the City", *Publications of the American Sociological Society*, 20, pp. 188-97.

INDEX